Access to the Genome
The Challenge to Equality

Access to the Genome
The Challenge to Equality

Maxwell J. Mehlman, J.D.
Jeffrey R. Botkin, M.D., M.P.H.

GEORGETOWN UNIVERSITY PRESS / WASHINGTON, D.C.

Georgetown University Press, Washington, D.C.   20007
© 1998 by Georgetown University Press. All rights reserved.
Printed in the United States of America.
10   9   8   7   6   5   4   3   2   1998

Library of Congress Cataloging-in-Publication Data

Mehlman, Maxwell J.
    Access to the genome : the challenge to equality / Maxwell J.
Mehlman, Jeffrey R. Botkin.
        p.   cm.
    Includes bibliographical references (p.   ) and index.
    1. Human genome.   2. Human Genome Project.   I. Botkin, Jeffrey R.
II. Title.
QH431.M366   1998
599.93'5—dc21
ISBN 0-87840-677-8 (cloth)
ISBN 0-87840-678-6 (pbk.)

                                                        97-37974

Contents

Acknowledgments

This book was written under a grant from the Ethical, Legal and Social Implications Branch of the National Center for Human Genome Research, National Institutes of Health.

The authors would like to extend their sincerest thanks to a number of individuals who were instrumental in the preparation of this book. The authors wish to thank Helena Rubinstein for her research and editorial assistance; Eric Juengst, former director of the Ethical, Legal and Social Implications Branch, for his encouragement; Jean Nash, Jay Jacobson, Leslie Francis, Peggy Battin, Ray Gesteland, and the Department of Pediatrics at the University of Utah for their support; Mellanee Kilpack and Jean Carter for their energy and efficiency; Walter Mantani for his illustration; Susan McIntosh for her research help; our editor, John Samples, for his editorial suggestions and faith in the enterprise; and our families for their understanding and support during this enterprise.

1
Introduction

Kate and Dan Britton planned and saved to be parents. Careful planning was necessary since it is an expensive enterprise and somewhat time-consuming. Both Kate and Dan are thirty-four years of age, which is the fashionable age to procreate within their social class. Having an unplanned baby would be highly embarrassing—social suicide really. One has to show personal and social responsibility in such important matters and lapses are not easily forgiven. The lower social classes continue to mate in the traditional way and have random children in random fashion with no efforts at quality control. Of course, contraceptives are available to all, but the less educated masses appear unable or unwilling to use them consistently. Accepting a child that is simply a chance combination of parental genes is unacceptable—like choosing a car by tossing a dart at the classified ads.

The process that Kate and Dan used is fairly routine. Kate had undergone "ovary hyperstimulation" over a three-month period through use of a variety of hormones. To spare repeated injections, a small capsule was implanted in Kate's leg that released hormones into her system on a programmed cycle. The hormones "pumped up" her ovaries to yield several eggs per month for the three-month time period. Some women would choose four- and five-month hyperstimulations, but this would cost more than the Brittons wanted to spend at this point. Besides, the three-month routine could be repeated in several months if no acceptable embryos were created. The eggs were retrieved through a slender needle inserted into Kate's abdomen with only minor discomfort.

Once retrieved, the eggs were automatically scanned for obvious defects and then fertilized with Dan's sperm. The sperm, too, had been scanned for damaged cells and sorted to enrich the quality. A new sorter had been introduced recently that could detect a dye for the Y chromosome. The harmless dye enabled the selection of sperm to create either all boys or all girls from the harvested eggs. The Brittons knew they wanted a girl and were delighted that they would not have to waste embryos by discarding the males. All the fertilized eggs would produce females. This immediately doubled their choice of embryos. The tough part was choosing the embryo actually to be implanted.

The fertilized eggs were permitted to divide to a sixteen-cell stage at which point two cells were removed for analysis. The process is largely automated and quite rapid due to the volume of customers being served. Nevertheless, parents were welcome to watch as their embryo was being analyzed. Cameras attached to microscopes tracked each set of cells, and the images were transmitted back to the parental lounges so couples could follow the action on TV. Kate and Dan splurged on a private room where leather chairs recline before a large, high-definition screen. Canned narration explained each event in simple terms as the cells moved down the disassembly line.

The amber cells appeared on the screen bathed in a clear, blue-tinged fluid, magnified to look like slightly irregular basketballs. (The colors were added to the image by computer following psychological research that identified the most soothing combination.) Each was held firmly to the tip of a microscopic glass tube by suction. Another microscopic glass tube, thinner yet, pierced the cell and extracted the entire nucleus, wherein lies the DNA—the genetic material. The rest of the cell, comprised of the cytoplasm (the gelatinous fluid around the nucleus) and the outer cell membrane, was sucked into a separate tube. This was really all that could be seen of the actual process since the rest was biochemical and genetic analysis. However, the images of the cells were interspersed with images of people in what looked like space suits moving about a laboratory. The laboratory was gleaming white with rows of complicated looking machines and video displays scrolling columns of numbers. The narrator explained that the DNA and other chemical analyses being done on the cells were so sensitive that droplets of saliva or microscopic flakes of skin from laboratory scientists could destroy the accuracy of the results—thus the need for the space suits. The cleanliness and the attentiveness of the scientists was very reassuring. Kate had read somewhere that the scientists being portrayed were fake and that the entire process was so automated that there was no need for scientists to be directly involved. She didn't believe it. These scientists looked real to Kate and, anyhow, they certainly wouldn't charge so much if the whole thing was done by machine! The results of each analysis were displayed on the screen for the parents. At that point, no attempt was made to explain the numbers and the technical names that went along with them, but it was reassuring that some results were obtained and that the scientists were smart enough to make sense of it all.

The cytoplasm from each cell was checked for enzyme, lipid, and carbohydrate content since these patterns strongly influence how the genetic material functions. (Overlooking the role of the cytoplasm had been a problem with this technology for decades. Now the predictive power of the combined genetic/cytoplasm analysis was really quite good.) The DNA of the cells was sequenced at 185 key sites that control or influence a wide range of diseases and physical and mental characteristics, ranging from seventeen different causes of mental retardation to shyness and skin color. More sites could be analyzed for a price, but most couples had found that the standard package already produced more choices than they could handle. Surveys had shown that most couples only wanted to choose among about twenty characteristics, but a number of uncommon diseases had been included in the screen, and tests for a variety of other physical and mental "traits" had been added to the package by the procreator profession for liability reasons. Experience had shown that people wanted some choice and some mystery, but they also did not want any rude surprises.

Of course the Britton's choice was not limitless. After all, their embryos contained a combination of their genes. Kate produced fifteen eggs so the couple now had a choice of fifteen female embryos, each with its own unique combination of their genes and the cytoplasmic factors in Kate's eggs. The ability to predict what each combination of genes and cytoplasm would produce was enhanced by the correlation with Dan and Kate's own genetic patterns. They, too, had their DNA analyzed at the 185 sites, and this genetic information was correlated with data about their physical and mental characteristics. Understanding how certain genes functioned in Dan, for example, would greatly assist in the prediction of how these same genes might influence Dan's daughter. The information from the analyses of each embryo, together with Dan and Kate's analyses, were fed into a computer program that produced a three-dimensional image of each potential child at any age selected by the viewer. The company emphasized that these were only rough predictions, with no guarantees, but Kate and Dan found the images startling.

The Brittons took some pride in their modest expectations. They wanted their daughter to be tall and slender, like Kate (and like virtually all of the women in their social class). Hair color was also to be light, although a sandy brown was acceptable since Dan had brown hair. (The very tall, white-haired children emerging from experimental labs in China had created an immediate demand for such extraordinary children here, but the Brittons thought this too strong a "statement." Anyhow, gene insertions were necessary for this effect, something for which the Brittons did not want to pay.) They wanted a child with light skin and an athletic build, bright but not necessarily brilliant (Again, they had to take what they could get and the Brittons were not brilliant, although a combination of genes from each might surpass their individual qualities.) The personality traits were of prime importance. Their ideal daughter was to be modest but not shy, kind, energetic, and diligent in her work. Kate and Dan realized that such traits were not purely inherited and that they had their work to do as parents. Nevertheless, they wanted the raw material to be right.

nature v. nurture

A computer program selected an embryo that most closely matched parental requests. The computer selected embryo #8 for Kate and Dan. This was odd, they thought, for in *their* review of the reports, embryo #13 seemed a much closer match for their key characteristics. But there it was, in red italics toward the end of the trait list for embryo #13: "Diabetes risk." This meant that a child from this embryo would have a greater than one in twenty risk of diabetes—the standard risk cut-off for the computer program. Diabetes was certainly a manageable condition, but it was a glaring flaw in this otherwise beautiful embryo.

The procreation counselor outlined the choices for the Brittons. (1) Choose embryo #8, not ideal but a perfectly acceptable embryo. (2) Choose embryo #13 and hope that diabetes does not develop or, if worse comes to worse, manage it with the usual metabolic implants. (The implants "cured" the diabetes, but they had to be implanted yearly, and diabetics still had to keep an eye on the condition to make sure the implants were working appropriately.) (3) Choose #13 and target the diabetes susceptibility gene with a replacement gene. Choice #3 would eliminate the risk but was only 90 percent effective. More problematic, though, was its expense—much more than the Brittons had anticipated for procreation.

The choice of an embryo is a monumental event, and couples agonize over the images and lists of traits. Divorce for irreconcilable differences over selection is not uncommon—both at the time of selection and later if the child demonstrates undesirable qualities of one of the parents. Kate was in tears for a week as they struggled with the choice. Dan was distracted and edgy, eventually deciding to stay home for a few days to sort it out. What were their priorities in life? What could be more important than one's children and one's genetic legacy to future generations? What was one's obligation to society with respect to the quality of one's children? What would others think about a choice to knowingly risk future illness? As their anguish and frantic anxiety began to ease, the answer became increasingly clear.

The gene implant was scheduled for the following week. The Brittons would have to take a second mortgage on their townhouse and probably say goodbye to their dreams of a vacation home in Aspen, but after the replacement gene was inserted, their perfect child would be born from embryo #13. Their love then would be unconditional and complete.

∗   ∗   ∗   ∗   ∗

Kate and Dan's story sounds like science fiction. Yet each step in this futuristic embryo selection process is either available now or under active development. Egg retrieval, in vitro fertilization, and embryo testing for genetic abnormalities are all commercially available. The ability to test for genes that influence physical and behavioral characteristics is not available, and there is controversy over the degree to which these complex traits will ever be amenable to prenatal genetic diagnosis. Nevertheless, there is little doubt that these characteristics have a significant genetic basis, thus opening the possibility that the ability to test for them will become available, at least in a rough manner, in the not-too-distant future. In fact, it is entirely possible that the scenario above will be a reality within twenty years.

These advances in genetic and biologic knowledge raise fundamental questions. What kinds of choices will people make when we gain creative control over life on the planet? Perhaps the question reflects simplistic hubris—it is likely that *complete* control will elude science for quite some time, quite possibly forever. The history of science suggests that things are more complex than we imagine, and this likely will prove as true for biology as it has for physics. Yet greater and greater measures of control over life's processes will be gained in the coming decades. Bacteria will be designed to manufacture many drugs and a host of industrial products and to eat up the toxic mess we have left behind us. Plants will be fashioned for disease resistance, higher yields in established environments and modest yields in formerly hostile environments, and for greater freshness at the market.

Animals will be genetically bred for higher yields for products, for human consumption, and for other valuable attributes such as blood and organs that can be transplanted to human bodies. Children will be designed for . . . for what? For good looks? For good health? For superior intelligence?

As genetic technology and other technologies relevant to human biology are developed, society will face a host of complex ethical, legal, and social issues. One of the most complex sets of issues involves our choices over who will have access to these technologies. The Brittons were able to engage in their embryo selection process because they were able to afford to. They were highly educated. Their social status motivated them to pursue these genetic techniques of procreation. But others—the "lower social classes" in our scenario—did not have children in this fashion, perhaps because they did not want to. More likely, it was because they could not afford to. If, indeed, diseases can be cured or prevented, and if embryos can be selected or enhanced, but the technologies are not available for all to have, how will our society respond? What does justice require, and what will political realities permit?

The issues surrounding how to distribute access to the new genetic technologies are the focus of this book. These questions and problems are not new—we struggle now with the fair allocation of vital health care resources in a society marked by great discrepancies in wealth. But genetic technologies will dramatically up the ante as the menu of powerful possibilities expands. With the possible exception of slavery, these technologies represent the most profound challenge to cherished notions of social equality ever encountered. Decisions over who will have access to what genetic technologies will likely determine the kind of society and political system that will prevail in the future.

Before discussing the social and moral issues raised by these questions, we will first provide some background information on the genetic revolution itself. In the next chapter, we describe the massive effort to decode the human genetic recipe. In Chapter Three, we describe the technologies that this effort is likely to produce and the potential benefits that they will provide. Then we will embark on a discussion of the social and ethical problems that these technologies pose and, finally, discuss potential solutions.

2
The Human Genome Project

David, the "bubble boy," was born only a few years too soon. He was born without a functioning immune system, which left him at risk of fatal infections from viruses and bacteria that the rest of us shrug off with minor discomforts (and, perhaps, an antibiotic). David was born in the 1970s when we had only antibiotics, intensive care units, and elaborate plastic bubbles that offered him temporary protection but no effective treatment and certainly no cure. Survival came at the cost of painful isolation, although David was given fame and the sympathy of the world in his tragic struggle.

Through the 1970s and 1980s medical care became increasingly complex, yet David's physicians had little choice but to use what Lewis Thomas called "halfway technology," that is, technology based on an incomplete understanding of disease.[1] This technology is halfway between a shaman's rattle and a sophisticated technology that flows from a fundamental understanding of biology. Halfway technologies are characterized by expensive and often burdensome interventions that are only partially effective—an accurate description of much of contemporary medical care. Kidney dialysis, coronary bypass surgery, and bone marrow transplantation are all complex and expensive forms of care, performed largely because we do not know enough to prevent or to provide truly effective treatments for the underlying conditions. In contrast, Lewis Thomas' example of a truly developed technology is vaccines. Based on a sufficient understanding of our immune system, these cheap and simple "shots" have drastically reduced the toll exacted by diseases that have plagued humans through history.

Current research in biology, and in genetics more specifically, promises to take us beyond halfway technology in the foreseeable future. David died in 1984, three years before Drs. French Anderson and Michael Blaese formally proposed the first attempt at gene therapy that could have targeted the rare disease that claimed David's life. On 14 September 1990 a four-year-old child, Ashanthi DeSilva, became the first human recipient of genetically engineered cells. Ashanthi has an inherited disorder termed "ADA deficiency," which is characterized by the lack of a single enzyme, resulting in a crippling of the entire immune system. While quite rare, the condition is ideal for gene therapy. The presence of even small amounts of the enzyme in the bloodstream reverses the immune deficiency, and the amount of enzyme does not need to be finely regulated with other body chemicals or systems. Ashanthi's gene therapy was accomplished by drawing a sample of her blood, inserting the gene for the enzyme into her own white blood cells, and then reinfusing the altered cells into her bloodstream. If the inserted gene would function in only a small fraction of her cells, Ashanthi would gain sufficient enzyme production to energize her immune system. It worked—at least to a modest degree. As her system became stronger, Ashanthi was released from her protective isolation and was free for the first time to go to school, to play with other children, and even to get her "shots" like other children. Yet at this first stage, the genetic therapy was not a genetic cure. White blood cells normally have a limited life span, and as cells altered with the added gene die, enzyme levels fall. Infusions of a new crop of altered cells must be administered every few months, at a total cost of approximately $250,000 per year.[2] In a second procedure designed to make the gene therapy permanent, the genes were inserted into Ashanthi's bone marrow "stem cells" that are the "immortal" parent cells of the blood system. If successful, all the white cells derived from the engineered stem cells would carry the gene through the course of her life, achieving a permanent cure. Whether this has been successful is not yet known.

These treatments are more than a clever approach to a single rare disease. They herald a new approach to medicine based on an understanding of normal and abnormal function at the cellular and molecular levels. Looking down the road, cumbersome "halfway" technologies will yield to elegant molecular approaches, just as iron lungs for children with polio have given way to vaccines for healthy infants. This transformation is being fueled by an enormous interest in research on genes and DNA.

Genes are often described as the biologic blueprints or recipes for life. DNA, in which the genes are found, carries genetic information from one generation to the next and encodes the basic plan that will fashion from single cells a human, a butterfly, or a fungus. Current research in biology is beginning to crack this DNA code, leading eventually to a fundamental understanding of many biologic processes.

There is every reason to believe that this understanding will permit control over many biologic processes, and biologic control will transform medicine, agriculture, animal husbandry, pharmaceutical production, and, perhaps, all other human endeavors that involve living organisms.

An understanding of human DNA certainly will be a vitally important key in understanding a host of human diseases. Cancers, in particular, are now being understood as "genetic diseases," since cancerous growths arise from either acquired or hereditary changes in cellular DNA. Once we know how altered DNA induces cancer development, elegant tools can be developed to prevent or treat malignant growths. The war on cancer will be much more effective once the tactics and defenses of the enemy are understood. Some recent work offers a glimpse at the potential of cancer-fighting tools to come.[3] In the not-so-distant future, brain tumors may be intentionally infected by viruses engineered to bind exclusively to the tumor surface but not to normal cells. Within the virus will be a gene, not found in normal human cells, that breaks down a normally benign drug into a toxic chemical. After the drug is administered to the patient, the inserted gene in the tumor cells will break down the drug, producing a toxic reaction in the brain tumor. The tumor will melt away as healthy cells remain unscathed. Treatment of cancer may become as easy as treatment of pneumonia is now.

Whether or not this type of therapy proves to be feasible, there is no doubt that our descendants will view our response to many diseases as crude at best. Cancer chemotherapy, radiation, and surgery are burdensome and often ineffective because of our lack of basic biologic knowledge. Autoimmune diseases, many blood diseases, heart disease, hypertension, metabolic diseases, many degenerative diseases, and, perhaps, a variety of psychiatric conditions will be exposed for effective attack when their basic biology is better understood.

DNA, GENES, AND CHROMOSOMES

In order to understand the genetic revolution that is now underway, it will be helpful to provide a brief overview of human genetics and the federally funded research endeavor called the Human Genome Project. (Interested readers can find more detail in several overviews

written for those without an extensive background in biology.⁴) An
individual's heredity is determined by the DNA that is provided
through the mother's egg and the father's sperm. The egg and sperm
each contain DNA packaged in twenty-three chromosomes. A new
embryo, made up of the DNA from the combined egg and sperm,
therefore contains forty-six chromosomes in each of its cells. These
forty-six chromosomes contain the entire human genetic code, called
the human "genome." As the embryo grows and develops into a fetus
and into an infant, each of the body's cells retains a copy of the forty-
six chromosomes.

A chromosome is a long stretch of DNA that is tightly coiled and
surrounded by certain proteins. DNA is composed of two strands of
sugar and phosphate molecules that are wound around each other in a
double helix configuration. Each of these backbone strands is studded
with a string of molecules called "nucleotides," which are composed
of one of four molecules—adenine, thymine, guanine, and cytosine.
Each nucleotide molecule is chemically bonded to another nucleotide
on the other strand. Together the two nucleotides make up a "base
pair." Adenine (A) always pairs with thymine (T), and guanine (G)
bonds with cytosine (C). If we were to read along a sequence of nucleotides on one strand of DNA, we might read a sequence such as
"ACCGTGACTG." Once the bases on one strand are known, the
sequence on the other strand can be deduced quickly due to the constant pairing of A with T and C with G.

The sequence of base pairs carries the genetic information contained within the DNA molecule. There are three billion base pairs in
the human genome. The biologic machinery of the cell translates a certain sequence of base pairs into a sequence of another set of molecules
called RNA. RNA, in turn, is translated into a sequence of amino acids,
which are the building blocks of proteins. Through this mechanism, a
specific sequence of DNA will produce, or "code for," a specific protein. Proteins do the yeoman's work for the cell by providing the raw
materials for cellular structures and by creating enzymes that facilitate
the tens of thousands of chemical reactions that are essential to life.
Since DNA controls the type, amount, and sequence of proteins produced in a cell, it earns the title of "mastermolecule" or "blueprint for
life." This same basic flow of information from DNA to protein is utilized by the vast majority of life forms on earth, from the simplest bacteria to the most complex organisms.

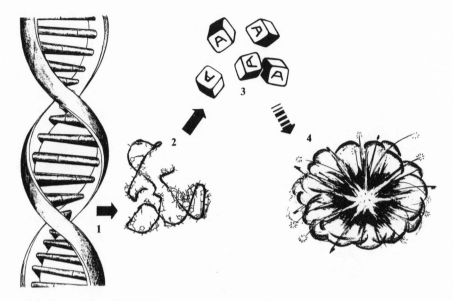

FIGURE. 1) DNA base pairs translate to 2) RNA, which translates to 3) amino acids, the building blocks of proteins, which create enzymes that generate 4) chemical reaction.

This rather complicated process can be understood by referring to the simplified diagram above.

So far, we have not mentioned genes. A gene is the stretch of DNA base pairs that codes for the production of a specific protein. There are estimated to be fifty thousand to one hundred thousand genes within the full complement of DNA that makes up the human genome. Only a small percentage of human DNA codes for protein. The existence of the noncoding DNA poses a difficult problem for scientists in identifying functional genes within the enormously long stretches of DNA. Once a gene has been identified in a general region of a chromosome, the DNA must be analyzed bit by bit to detect the telltale sequences that contain start or stop signals for active protein coding. At the present time, the function, if any, of the noncoding DNA in the genome, sometimes referred to as "junk" DNA, is unknown. However, it is likely that junk DNA will prove to have important functions in the cell, or perhaps it had important functions in the past, during the evolution of our species.

A normal, functional gene can be altered by a variety of factors, including errors in copying from one cell to another by the internal cell machinery, and external factors, such as radiation or chemical exposures. While many alterations in the DNA sequence will prove to be harmless, some will lead to the production of proteins that cause illness or deformity. These abnormal gene sequences are termed "mutations." Genetic mutations that are inherited from parents will be present in all the cells of the body. Given the number of base pairs in a single gene, many different mutations are possible in any given gene. For example, any one of six hundred different mutations in a given gene will produce the disease cystic fibrosis. The type of mutation in a gene often will have consequences for the individual who carries the mutation. Some mutations will cause the protein to be completely nonfunctional, while others may permit varying degrees of residual activity. This may mean that some individuals with a genetic disease are much more severely affected than are others with the same condition.

Since each of us has two of each chromosome, and therefore two of each gene, mutations in only one gene may produce no ill effects. As long as the person has one functional gene, they may not be adversely affected. Sickle cell disease is a good example. Sickle cell disease is caused by a mutation in the genes for hemoglobin—the oxygen-carrying molecule in red blood cells. Two abnormal copies of the gene (one from each parent) lead to sickle cell disease, which is characterized by episodes of illness in which the individual's red cells become deformed ("sickle" shaped), causing blockage of blood vessels throughout the body. The short-term effect of the disease is acute, severe pain; the long-term effects are disability and an early death. However, individuals who carry only one abnormal copy of the gene are entirely healthy. The abnormal hemoglobin can be detected in their blood, but the concentration of the abnormal protein is not sufficient to produce disease. Conditions such as sickle cell that require two abnormal copies of a gene to produce disease are termed "recessive" conditions.

For some conditions, however, one abnormal copy is sufficient to produce illness. A gene may produce enough abnormal protein at one of the two gene copies to lead to impaired tissues. Imagine building a home in which one of every two pieces of lumber has a large crack. The resulting structure would be seriously flawed and prone to catastrophic failure. Conditions in which only one abnormal gene is required to produce disease are termed "dominant" conditions.

A genetic mutation may only affect a small number of tissues in which the gene is active. Cystic fibrosis (CF), for example, is a recessive condition that results from an abnormal protein that regulates the transport of chloride across cell membranes. Individuals with CF have the abnormal genes in all of their cells, but the impaired chloride transport seems to be important primarily in lung mucus production and in pancreas function. As a result, CF is characterized by progressive lung disease and poor digestion from pancreatic insufficiency.

For some genetic diseases, and sickle cell disease is a good example, it is possible to detect the abnormal protein or its associated physical manifestations in the body. The analysis of the abnormal hemoglobin blood cell in sickle cell patients led to the detection of the mutated hemoglobin gene. Unfortunately, we do not know the primary biochemical abnormality for a wide range of diseases which result, at least in part, from genetic factors. Without an abnormal protein in hand, scientists cannot work backwards to detect the responsible gene.

Increasingly, scientists are searching for the abnormal gene first and then elucidating the function of the gene and its associated protein. The understanding of human disease (and normal human function) will be greatly enhanced if we understand the location and, subsequently, the function of all of the genes within the human genome. This is the rationale for the fifteen-year, multibillion-dollar effort called the Human Genome Project.

THE HUMAN GENOME PROJECT

In the 1980s, it was becoming increasingly apparent to many scientists that an understanding of basic biology would be greatly enhanced if the detailed structure of DNA was understood. In addition, technology was emerging that gave scientists confidence that such a massive undertaking could be successful. The debate in the scientific community over whether to pursue this goal led to key reports in 1988 by the National Research Council of the National Academy of Sciences[5] and the Congressional Office of Technology Assessment.[6] Both supported the idea, and, despite a few voices of concern in the scientific community, Congress allocated funding for the creation of what became known as the Human Genome Project.

To oversee the work, a new organization was established at the National Institutes of Health called the National Center for Human Genome Research. A smaller portion of the project was delegated to the U.S. Department of Energy, based on that agency's experience with genetic research related to radiation effects and its background in computer science. James D. Watson, Ph.D., the famous codiscoverer of the DNA double helix, was recruited to direct the NIH effort. Watson has been enthusiastic about the prospect of seeing scientific knowledge expand from the elucidation of the basic structure of DNA (in 1953) to a complete catalogue of the three billion base pairs in the human genome—all to be achieved within the lifetime of one scientist. According to Watson:

> A more important set of instruction books will never be found by human beings. When finally interpreted, the genetic messages encoded within our DNA molecules will provide the ultimate answers to the chemical underpinnings of human existence. They will not only help us understand how we function as healthy human beings but will also explain, at the chemical level, the role of genetic factors in a multitude of diseases—such as cancer, Alzheimer's disease and schizophrenia—that diminish the individual lives of so many millions of people.[7]

It bears some emphasis that the scientific work of deciphering the basic structure of human DNA has been going on for decades and would have proceeded inevitably towards the same goals without the organized effort of the Human Genome Project. The advantage of the Human Genome Project is that it has attracted extra funding to the work, raised the profile of the effort within the scientific and lay communities, and provided elements of organization and cooperation that would not have occurred with individual scientists pursuing projects based on their personal interests. An additional advantage of a high-profile Human Genome Project has been the recognition that we are ill-prepared as a society to deal with many of the complex problems that will arise from advances in genetic technology. The eugenics movement in the United States and other Western countries, and the support that eugenics philosophy gave to the horrors of National Socialism in Germany, have made many people appropriately sensitive to the potential abuses of genetic science. Beyond abuses, there are basic problems in the application of genetic knowledge in medicine and society, including issues of the benefits and harms of testing and screening, issues of privacy and confidentiality, issues of regulation, and issues of justice in access to these powerful new tools. Three to five percent of the Human Genome Project funds have been allocated to conduct research and education on the ethical, legal, and social implications of human genetic research. To date, this "ELSI Program" as it is called, has invested more than forty million dollars toward these goals.

The goal of the Human Genome Project is to "map" and "sequence" the entire human genome. The concept of sequencing is the more straightforward, although the more difficult from a technical standpoint. As noted above, DNA is composed of a long sequence of base pair molecules (A's, T's, C's, and G's) linked to a helical backbone. The goal of sequencing the genome is to identify, in the proper order, the three billion base pairs that make up human DNA. Through the process of compiling this information, the telltale sequences that signify genes can be located and the exact sequences of all the genes revealed. So the process of sequencing will provide the real information that will be useful—the location and base-pair structure of all of the genes that code for humans.

An often asked question is *whose* genome will be sequenced in this project? The answer is that while a single or limited number of individuals may be used as a reference standard for sequencing efforts by many laboratories, the knowledge gained will be applicable to us all. As noted, there is considerable variation from one individual to the next in the sequence of their genes and in the sequences of "junk DNA" between genes. Nevertheless, it is assumed that the basic organization of human DNA is the same in all people. Identifying where the genes lie and what their functions are in one individual will reveal the location of the same genes in virtually all other individuals. Here the analogy of human anatomy is useful again. We all have some variation in the structure of our kidneys, brains, hearts, and other organs, but a detailed anatomic study of any of us would yield useful information about how all of us are built. Subsequent research will be necessary to determine whether the variations that are seen between individuals represent abnormalities or only benign variants.

In fact, the basic structure of DNA is similar in organisms that are not too distantly related to humans from an evolutionary perspective. One of the primary goals of the Human Genome Project, for example, is to map and sequence the genome of the mouse. The mouse, a fellow mammal, possesses DNA and a genetic structure that are remarkably similar to those of humans. Further, mice reproduce rapidly, producing large numbers of progeny, and, unlike humans, mice can be mated and manipulated as investigators see fit. Studying the sequence of the mouse genome can provide important information about the human genome.

The other goal of the Human Genome Project is to "map" the human genome. The basic idea of a genome map is that it establishes landmarks throughout the genome that can be used as reference points to locate genes in their vicinity. These genetic landmarks are unique sequences of DNA that can be identified by the use of genetic tests called "probes." A map is created by identifying sequences, called "markers," that are positioned along each of the forty-six chromosomes. Once a map has been constructed, each gene on a chromosome can be located in terms of its position relative to a marker.

To illustrate this more clearly, consider a large family that is affected by a genetic disease Z that typically affects individuals in their 50s. Imagine that the gene for disease Z has not been identified and, further, that scientists do not have any idea how the gene controlling disease Z functions. The search for the gene can begin with blood samples taken from affected and unaffected family members. The DNA from family members is laboriously screened with probes for markers scattered throughout the genome. In a process called "linkage analysis," family members with and without the disease are checked to see if they have the same markers. The investigators are looking for a marker found only in people with disease Z; in other words, a marker that seems to travel with the disease. The investigators will know they have hit paydirt when a marker is consistently carried by those who are affected with disease Z but is not present in unaffected family members. The presumption is that the marker is "linked" to the responsible gene, in the sense that the marker is physically close to the gene on the DNA molecule and the two travel together through the generations of the family. A marker can serve as a very useful tool for "presymptomatic" diagnosis of the condition. For example, if a young adult carries the same marker as affected family members, it is highly likely that the individual carries the gene and will develop disease Z in later years.

Two aspects of this approach are worth emphasizing. First, the markers themselves are important, not because of their own genetic function, but because they serve as reference points for nearby genes. Most of the markers are located in the so-called "junk" or nonfunctional DNA regions. Second, markers and linkage analyses allow individuals who carry the gene to be identified before the gene is located or its function understood. Testing for Huntington's disease is a good example. Huntington's disease is a dominant hereditary condition that leads to progressive neurologic deterioration beginning at about forty to fifty years of age. A marker for the gene was found on chromosome 4 in 1983.[8] This marker enabled clinical testing to be offered to people at risk for the disease. However, it was not until 1993 that the gene responsible for Huntington's disease was found and copied, or "cloned," so that it could be detected directly in genetic tests.[9]

Once a gene has been reliably linked with a marker, testing can be performed to determine whether an individual has inherited the mutation that causes the disease. However, genetic research is providing the ability to diagnose genetic conditions well in advance of our ability to effectively treat or prevent disease in those who are identified as at risk. The linkage of a gene to a marker is key to enabling scientists to isolate the gene itself. Even with the gene in hand, though, much more work remains to be done to translate this knowledge into more effective medical care.

Thus far, we have discussed human genetics largely in the context of disorders, such as sickle cell disease and cystic fibrosis, that are associated with a mutation in a single gene. While single gene disorders represent a very important category of diseases, it would be misleading to think of genetic conditions in these terms alone. First, it would be simplistic to think of a genetic mutation as *the cause* of a genetic disease. Genes function in a complex interaction with other genes, and with the environment within and beyond the cell. Even with single gene conditions like CF, the presence of other modifying genes and environmental factors can play a significant role in determining the severity of the condition. Some children with CF still die in the first years of life, while others with the condition are living into their 40s. Similarly, phenylketonuria (PKU), which causes mental retardation, is due to a single gene defect that impairs an individual's ability to metabolize one amino acid, phenylalanine, into another amino acid, tyrosine. Mental retardation caused by PKU can be prevented if a diet is provided that contains no phenylalanine. In this case, a genetic disease can be prevented by an environmental change.

At a second level of complexity, there are conditions that are termed *polygenic* and *multifactorial*. Polygenic conditions are conditions that require the effects of more than one gene to produce disease. Multifactorial means there must be both genetic and environmental factors involved. In practice, it may be difficult to distinguish between polygenic and multifactorial conditions. At any rate, these categories include some of the major causes of disease and death in humans, including atherosclerotic heart disease, diabetes, asthma, hypertension, manic depressive illness, and schizophrenia. Family studies suggest that there are significant genetic components to these diseases, yet they are clearly not inherited in a simple dominant or recessive fashion. While it is likely that there are single genes that play a large role in the development of these conditions, there must be other genes and/ or environmental conditions that have strong influences as well. Obviously this has great importance for the application of tests for genes that will be found to be associated with these conditions. Detecting a gene associated with schizophrenia in an individual, for example, may mean only that she is more likely than the general population to develop the disease. She may still have a good chance of never being affected. Information from genetic testing may be of little real use to the individual unless testing can detect the other genes that contribute to the illness, or unless clear environmental factors can be identified. The important point here is that having the gene for a condition is not the same as having the condition.

The Human Genome Project and the great strides that have been taken in molecular biology in recent years have produced an enormous excitement in the scientific community. Originally projected to be completed in 2005, the Human Genome Project may be ahead of schedule. A complete map of the human genome was completed in 1994, although additional work is necessary to identify more markers. Attention is now switching to sequencing. A proposal has been floated recently to reorganize the effort to sequence the entire human genome.[10] This proposal would reallocate funding from developing new sequencing techniques and invest it in current sequencing technology on a massive scale. Advocates claim that this would enable the human genome to be sequenced more cheaply than anticipated and to be completed by the year 2001—a full four years ahead of schedule. The ability to identify genes associated with diseases will be accelerated accordingly.

While the public has been informed of many of the genetic discoveries as they have rolled out of the labs, the real results remain in the future. For every Ashanthi DeSilva who is offered a "miracle cure" now, there will be hundreds of thousands in the coming decades who will benefit from this knowledge. If carefully controlled, the knowledge and technology that are emerging will herald a fundamental change in our relationship with the biologic world. The potential benefits from these technologies are described more fully in the next chapter.

book was published 1998

3

The Practical Applications of Human Genetic Technology

Biological knowledge affects health once it is translated into practical tools. The knowledge obtained from the Human Genome Project will have tremendous implications for the future of medicine as we develop genetic technologies that enable us to prevent and treat a wide range of clinical conditions.

It is important to discuss briefly what we mean by "genetic technology." In the near future, much of the technology emerging from the Human Genome Project will focus directly on the genes themselves. A technique that uses short strands of genetic material, called DNA probes, will be used to detect normal and abnormal genes for diagnostic and screening purposes, and gene therapy will be used to transfer genes into cells to repair, alter, or enhance their function. But as our knowledge becomes more sophisticated, diagnostic techniques may focus on detecting the presence or absence of gene products, such as proteins, rather than the gene itself, and therapies for genetic disorders will be based on a manipulation of a cell's structure or metabolic functions rather than its DNA. Of course, this is exactly what medicine is doing now for many genetic diseases, although not necessarily very effectively. We diagnose sickle cell anemia by analyzing hemoglobin rather than DNA, and we treat cystic fibrosis with antibiotics and pancreatic enzymes rather than with altered DNA (although gene therapy trials are underway for CF). The point is that genetic or DNA-based technology is not the ultimate goal of genetic research. The goal is an understanding of biology that will yield DNA-based tools, as well as tools based on other aspects of the life process. It is likely that the first generation of genetic technologies will be DNA-based. But by the term "genetic technologies," we also mean to refer to later, more powerful interventions that stem indirectly from our new understanding of the human genome.

Genetic technology in humans will yield three major practical applications: screening and diagnostic technology, gene therapy, and genetic enhancement. The development of a genetic map will make genetic testing and screening available for a rapidly expanding list of conditions. Thus, genetic diagnostics will be the first significant application of DNA technology. Genetic therapies, such as the one used for Ashanthi DeSilva, are now emerging experimentally and have great promise for a number of medical interventions. We also will speculate on the possible application of genetic technology for the enhancement of "normal" functions. Genetic enhancement is not on the immediate horizon, but we have little doubt that such applications will be feasible in the next twenty to thirty years, and will be widely desired.

GENETIC TESTING AND SCREENING

A common application of genetic technology will be the diagnosis of patients who are suffering from disease. Diseases amenable to genetic testing will include heritable conditions, acquired genetic conditions, and diseases caused by infectious agents.

Patients with Symptoms of Disease

Patients with a constellation of symptoms often undergo testing to identify the cause. Consider a six-month-old infant who is described by her parents as "floppy," meaning that she lacks the active muscle tone of other infants and cannot sit or roll over. There are a number of potential causes for such a condition in an infant, including several hereditary disorders of muscle and nerve tissue. At the present time, a diagnosis is based upon observation of the child's condition, supported by the results of laboratory tests on blood and tissue samples. For these patients, DNA-based diagnostic tests will play an important role in detecting genetic causes of their diseases. Detecting a genetic mutation will help confirm the medical diagnosis so that treatment decisions can be made with greater certainty. Identifying the specific mutation also may give the physicians information about the likely severity of the condition. Genetic confirmation of the diagnosis will enable the parents to make a decision about whether to have a future pregnancy, and, if a pregnancy is pursued, DNA testing can be used to determine whether the fetus bears the mutation in question. Genetic testing has value in this circumstance even though there may be no specific treatment for the condition: parents can be prepared emotionally and financially for the child's condition, or they may elect to abort the fetus. Of course, when effective interventions are developed in the future for such conditions, the value of genetic diagnosis will be that much greater.

For some genetic conditions, we have the ability to test for gene products or for characteristic physical signs of the abnormal gene function. It is likely that DNA-based genetic tests will not entirely circumvent these conventional tools in the near future. Cystic fibrosis again is a relevant example. The gene associated with CF has been located and more than three hundred different mutations have been described, although only a few mutations account for the majority of the cases. This large number of mutations means that there is no single DNA test that can be used to test a child who presents with suspicious symptoms of CF. Since a DNA test only detects a single mutation, more than six hundred tests would be needed to assure that the child does not have a CF mutation. For this reason, clinicians still rely on the traditional "sweat test," measure of chloride levels, found to be elevated in the sweat of persons with CF.

DNA-based diagnostic tools for symptomatic patients will be most useful in children in whom a heritable condition is suspected, since most inherited illnesses appear in childhood. However, DNA testing will become increasingly useful in a number of clinical circumstances in the adult population, including in such conditions as adult onset polycystic kidney disease, Huntington's disease, hemochromatosis (a disease in which iron builds in the body and causes death from heart disease), or alpha1-antitrypsin deficiency (causing breathing difficulties which can be fatal).

Acquired Somatic Defects

As discussed in Chapter Two, genetic abnormalities that produce disease, instead of being inherited, can be acquired during an individual's lifetime. Most cases of cancer fall into this category. Current thinking supports a "two hit" hypothesis that suggests that two genetic alterations are necessary in many circumstances to produce the genetic changes in a cell that lead to cancerous growth. Since each cell has two of each gene (one from each parent), "two hits" may be required to cause abnormalities in both genes, which then leaves the cell without the function of either copy. In some families with a hereditary predisposition to cancer, one copy of a key gene may be dysfunctional from birth, leaving a carrier of this gene susceptible to cancer if a second "hit" affects the other copy of the gene. "Hits" may come from errors in cell replication, radiation, chemicals, viruses, or other environmental influences.

Genetic technology may be useful in characterizing these acquired genetic abnormalities in symptomatic individuals because it may provide important information about the likely course of the disease. Cancer tumor cells would be analyzed for their genetic structure. There is some evidence that certain genetic features of breast tumors indicate a higher likelihood that the cancer will metastasize, or spread, early on.[1] In the future, this type of information may be a key factor in deciding how aggressively these tumors should be treated. In the more distant future, knowledge of the genetic structure of the tumor will allow specific therapies to be developed.

Infectious Diseases

Genetic technology also will become increasingly important in diagnosing infectious diseases.[2] Bacteria, viruses, fungi, and parasites each contain unique sequences of DNA that can be identified. At present, a frustrating limitation in the care of patients suspected to have significant infections is the need to wait twenty-four to forty-eight hours (and even longer for viruses and fungi) for the infectious agent to be cultured and grow so that there is enough of the agent to be identified. This delay often requires patients to be placed on broad-spectrum antibiotics prior to the identification of the infectious organism. Greater use of genetic tests, which can be automated, will speed the identification of infectious agents in patients, and allow more judicious use of antibiotics to avoid producing disease-resistant strains of infectious organisms.

Genetic Testing for Susceptibility to Future Illness

Susceptibility testing or screening is intended to identify individuals who are at risk for *future* illness. *Testing* is conducted on individuals who are at a higher than average risk of carrying the genetic mutation in question. *Screening* is performed on a wider population, including those who are at low risk. For some conditions, identifying a genetic mutation will enable specific preventive measures to be taken. In other cases, there may be no treatments or preventions available, but individuals may want to know the information in order to make life decisions, including reproductive choices. In the general category of genetic risk analysis, we will discuss three applications of genetic tests: late onset conditions, prenatal screening and testing, and newborn screening.

Late-onset conditions

One of the most significant applications of genetic technology will be to predict future illness or susceptibility to future illness in children and adults. Predictive testing would be appropriate for rare conditions in which a family history of the disease alerts the family and physicians to the need for testing. This circumstance would arise for such conditions as adult polycystic kidney disease, familial polyposis (which can lead to the development of colon cancer), Friedreich's ataxia (a movement disorder that destroys balance), G6PD deficiency (a blood disorder that causes anemia), Charcot-Marie-Tooth disease (a neurological condition that limits mobility), some breast cancers, and Huntington's disease. For some of these conditions, preventive measures may avert or ameliorate serious disease. Familial polyposis, for example, is managed by frequent colonoscopies and removal of the colon, if necessary. Knowledge of G6PD deficiency can alert the family to avoid certain drugs or foods that trigger a sudden breakdown in blood cells. For other conditions, however, there are no effective measures or treatments that can prevent or significantly ameliorate the disease, although testing in these circumstances allows individuals to plan their lives with a better prediction of future health status.

Once specific genes have been isolated for the more common conditions, it will be possible to perform "population screening" for these abnormalities. As noted earlier, in population screening, individuals are tested who do not necessarily have a family history of the disease. For example, two genes, called BRCA1 and BRCA2 have been identified.[3] Mutations in BRCA1 confer a lifetime risk of approximately 85 percent for breast cancer and 30 percent for ovarian cancer. BRCA2 also confers a high risk of breast cancer, but the risk of ovarian cancer does not appear to be highly elevated. Roughly between 1 in 200 and 1 in 400 women in the general population carries a mutation in one of these genes. However, as with cystic fibrosis, hundreds of different mutations are being found in these genes, so no simple test will be available for use in the general population. Subgroups of the population present a different story. Recent research has shown that only two mutations are present in as many as 1 or 2 percent of Ashkenazi Jews in the United States. (Ashkenazi Jews have an Eastern European background and comprise 90 percent of the Jews in the United States). Therefore, it has become possible to screen the population of Ashkenazi Jews for BRCA1 and BRCA2 mutations with a relatively simple set of genetic tests. Whether such testing is actually beneficial to women has not yet been determined.

The primary causes of illness and death in the United States, aside from trauma from such events as accidents and gunshot wounds, are heart disease, cancers of the breast, lung, and colon, diabetes, and stroke. Major illness is also caused by diseases of the central nervous system, including Alzheimer's disease, bipolar illness, schizophrenia, addictive disorders, and autoimmune disorders, including rheumatoid arthritis and lupus erythematosus. Many of these conditions have major genetic components. These conditions are sufficiently common that population-based genetic screening to detect persons at risk for these diseases might seem worthwhile.

Numerous difficulties would arise in providing this type of genetic testing. These are multifactorial illnesses that involve a complex interaction between genetic and nongenetic or "environmental" factors. Defining an individual's genetic status would provide only a limited prediction about whether or not the individual will become ill with the condition. Cholesterol screening provides an example. The efficacy of cholesterol screening is questionable, for most people with high levels never get heart disease, and some people with low levels will get heart disease anyway.

The fact that environmental factors interact with genetic factors would not significantly undermine the value of genetic testing if these interactions were understood and could be factored into predictions of risk. But this is not the case at present, where the interplay between genes and the environment is only beginning to be unraveled. Moreover, even if we knew what specific environmental hazards susceptible individuals needed to avoid, we have only a limited ability to change individual behavior. For example, it has been suggested that if children could be tested to determine who was more likely genetically to be affected with alcoholism or lung cancer secondary to smoking, we could influence specific teenagers to avoid cigarettes and alcohol. Given the independent nature of adolescents, however, it is highly questionable that this targeted influence would be successful. We also may be doing a serious disservice if individuals whose test results showed that they were at low genetic risk for conditions such as alco-holism or lung cancer incorrectly assumed that they could smoke and drink with impunity. These difficulties with population screening for multifactorial genetic ailments raise serious questions about whether, and under what conditions, it would ever be appropriate.

As with symptomatic testing, presymptomatic testing will also be possible for genetic defects in specific tissues. An example that is in development at the present time is the screening of stool for genes associated with colon cancers.[4] Current presymptomatic screening for colon cancer in the general population involves periodic testing for blood in the stool. However, blood in the stool may arise for many rea-sons other than cancer, and cancerous tumors do not always bleed. A DNA-based screening tool that would detect genetic mutations in cells shed in the stool could be a more accurate test for this disease. It would help avoid further unnecessary testing, and enable detection of cancers at an earlier and more curable stage of disease.

Prenatal screening and testing

Prenatal screening and testing is "presymptomatic" in the sense that the prospective parent is seeking information about the future health of a child. Virtually any genetic test could be applied prenatally. As with the testing commissioned by the Brittons in the futuristic sce-nario presented in Chapter One, cells separated from an embryo developing outside the woman's body can be tested to determine if the embryo would be affected by a genetically related disease. This approach now is being used for couples at risk for cystic fibrosis.[5] Women have their eggs harvested and fertilized by their husband's sperm *in vitro*, that is, in the laboratory. Embryos are grown to several cells and cells are removed for DNA analysis. Only embryos whose test results indicate they would not be affected with cystic fibrosis are introduced into the mother's uterus. As *in vitro* fertilization (IVF) tech-niques improve, this approach may become more widespread for cou-ples at risk for producing a child with a genetic condition. However, as we noted in Chapter Two, this approach need not be restricted to fami-lies with a history of genetic disease. Future parents who are interested in gaining greater control over reproduction and the genetic "quality" of their offspring also will be interested in this type of embryo testing technology.

The more traditional approach to prenatal testing, which occurs after the fetus begins to develop in the womb, also will be facilitated by genetic technology. The number of genetic conditions that will be amenable to prenatal diagnosis will expand significantly. The focus of present screening and testing is on relatively severe conditions, such as Down syndrome, spina bifida, Tay–Sachs disease, and, more recently, cystic fibrosis. As more genetic sites are mapped, we will be able to detect a broader range of conditions, from the severe to the relatively mild, and from early onset diseases that manifest themselves early in life, to late adult onset conditions that only emerge in adulthood, and sometimes, like Alzheimer's disease, not until relatively late in life. Genes that correlate with personality traits and general physical characteristics also may be detectable prenatally.

In addition to increasing the number of conditions that can be detected, genetic technology will improve the ease of testing. There have been preliminary successes in isolating fetal cells that circulate in the maternal bloodstream early in pregnancy. This offers the possibility of obtaining a simple maternal blood sample early in pregnancy and characterizing the fetal DNA with respect to a host of genetic factors. If this process could detect the thirty most common genetic diseases and attributes of interest, then a routine "genogram" of the fetus may become a standard obstetric practice. More specific tests will also be available for those with a family history of a particular condition. Currently, the primary justification for prenatal testing and screening is to offer the parents the choice of abortion if an affected fetus is detected. As prenatal medicine becomes more sophisticated, prenatal diagnosis will be performed in order to determine the need for prenatal treatments.

Newborn screening

Newborns are currently screened shortly after birth for a variety of congenital conditions, including enzyme deficiencies (PKU, maple syrup urine disease, homocystinuria, galactosemia), hormonal imbalances (hypothyroidism and congenital adrenal hyperplasia), and protein abnormalities (sickle cell disease and cystic fibrosis). These tests are performed by looking for the abnormalities in dried blood spots.

Genetic technology will enable newborns to be screened for a wide variety of genetic conditions, often with greater accuracy than current biochemical tests permit. As with prenatal screening, the spectrum of conditions for which testing will be available will expand tremendously to include milder conditions (e.g., asthma and lactose intolerance) and late onset conditions. It may be possible to tell new parents that their infant will have an increased risk of such conditions as diabetes, heart disease, breast cancer, alcoholism, or a variety of learning disabilities or personality traits.

Carrier Screening and Testing

As we discussed in Chapter Two, "carriers" refers to individuals who carry only one copy of a recessive disease gene and are therefore healthy, but who can pass the gene on to successive generations. For example, approximately 1 in 25 Caucasians in the U.S. population is a carrier of the cystic fibrosis gene. These individuals have no symptoms and no reason to suspect they are carriers for the condition. However, a carrier who has children with another carrier has a 25 percent risk with each pregnancy of having a child with CF. The purpose of carrier testing, therefore, is not to protect the health of the carrier, but rather to inform the individual about reproductive risks.

Carrier screening is an alternative to carrier testing. Carrier screening currently is available for such conditions as sickle cell disease, cystic fibrosis, Tay–Sachs disease, and the thalassemias (anemias caused by various abnormalities in blood production).

It generally seems preferable to perform carrier screening or testing during the reproductive years but before a pregnancy occurs. If the couple were both found to be carriers, then they could prevent the birth of an affected child by a number of means, including artificial insemination, surrogate pregnancy, egg donation, pregnancy and prenatal testing, or adoption. After a pregnancy has begun, carrier screening offers the couple a much more limited choice of terminating the pregnancy or continuing it with an affected child. At the present time, prenatal therapy for genetic conditions is not available beyond a very few conditions, although there is hope for great strides in the future. Of course, the vast majority of those who undergo carrier screening will be found not to be carriers. In this circumstance, carrier screening provides some reassurance to pregnant couples or to those contemplating a pregnancy. One further possibility for the future would be extensive carrier screening or testing of adolescents or preadolescents so that individual genetic profiles could be used in assuring that one's prospective spouse was genetically compatible.

In general, there is little purpose in conducting carrier screening for those who are beyond their reproductive years. Carrier screening of embryos and fetuses is possible, but unless parents are prepared to discard an embryo *in vitro* or abort a fetus because of its status as a carrier, there is little point in screening or testing at these ages. Newborns and young children may also be targets of this type of testing. Presumably those tested in childhood could use the information for marriage and reproductive decisions later in life, assuming that the information was remembered or was stored in a fashion that would permit retrieval years later.

GENE THERAPY

Gene therapy refers to the insertion of genetic material into a patient for therapeutic purposes. The experiments on Ashanthi DeSilva, described in Chapter Two, are a contemporary example of gene therapy. The potential for this technology is enormous. In the near future, we should be able to treat numerous conditions caused by the lack of production of a key protein. Recall that the basic function of genes is to produce proteins and that genetic mutations often produce nonfunctional proteins. The basic approach with gene therapy is to insert a gene into an individual's DNA that will produce the proper protein.

The thalassemias offer a good example of future prospects for this type of gene therapy. The thalassemias are a group of conditions characterized by the inadequate production of portions of the hemoglobin molecule—the molecule responsible for carrying oxygen in red blood cells. Chronic anemia results, which can be quite severe in some forms of the disease. Gene therapy would insert the gene for the hemoglobin subunits into the blood-forming cells in the patient's bone marrow. It is important to note that it is not necessary to insert genetic material into all of the target cells in order for the gene therapy to be successful. Even a low level of normal protein production can yield dramatic reduction in the severity of the condition. Much gene therapy experimentation currently is underway on conditions ranging from cystic fibrosis to familial hypercholesterolemia (an inherited condition of extraordinarily high cholesterol), Gaucher disease (a disease in which toxic levels of substances are stored in various organs, leading to early death), AIDS, lung cancer, kidney cancer, and hemophilia.[6]

As research in this area has expanded, the possibilities for even more creative uses of gene therapy have expanded as well. Gene therapy need not be restricted to hereditary diseases. It may also be used to create internal drug "factories" for a host of acquired medical conditions. It is not generally possible to supply a missing protein orally, since proteins are broken down by stomach acid and intestinal enzymes. Missing proteins are usually provided through intermittent intravenous or intramuscular injections: clotting factor for individuals with hemophilia and insulin for those with diabetes are examples. Therefore, the benefits of creating permanent internal supplies of therapeutic proteins are staggering.

In one pig experiment, investigators induced cells from the lining of blood vessels (endothelial cells) to coat the lining of an artificial Dacron blood vessel that was implanted in the pig.[7] The endothelial cells had been genetically altered to produce a specific marker protein. After the Dacron vessel was implanted into the pig, the endothelial cells continued to produce the protein and excrete it directly into the blood supply. One application of this system, discussed by the investigators, would be the insertion of the TPA gene into the blood vessels. TPA stands for "tissue plasminogen activator," a chemical that is injected into patients in order to dissolve blood clots and to treat and prevent heart attacks. Insertion of similarly altered endothelial cells to produce TPA could provide a patient at risk for a heart attack with a constant and permanent supply of TPA.

A third application of gene therapy involves destroying abnormal body tissues. In Chapter Two, we described an experiment in which a gene capable of converting a drug into a toxic chemical was inserted into brain tumor tissues in order to kill off the cancer cells. The ability to specifically target tumor tissues with genetic weapons could revolutionize cancer treatment.

Another excellent example is the current work at the NIH that is designed to augment the killing power of white blood cells directed against tumors.[8] In this research, white blood cells isolated from the patient's tumor are altered with the addition of a gene that produces "tumor necrosis factor," a substance that causes tumor cells to die. The white cells are then reinfused into the patient with the hope that the cells will infiltrate the tumor, secrete the necrosis factor, and destroy the tumor. These experiments have been dramatically effective in a few cases to date. A realistic possibility is the development of a number of powerful biologic agents that will home in on and destroy tumors in a much more selective fashion than our current tools of surgery and chemotherapy.

The potential benefits of gene therapy extend well beyond cancer and hereditary diseases. For instance, a person's own cells might be enhanced genetically to improve the cells' ability to repair damaged bones, nerves, muscle, skin, and cartilage. Another example where gene therapy would be useful is in the treatment of heart disease, the single largest killer in the United States. A relatively new method used to repair clogged coronary arteries is angioplasty, a procedure in which a small balloon is inflated inside the artery to expand it and increase the blood flow. But a significant problem with this method is the growth of smooth muscle tissue after the angioplasty that reblocks the vessel. In a technique similar to the brain tumor experiments described above, gene therapy has been used in pigs to destroy smooth muscle tissue in coronary arteries after angioplasty.[9] Someday this technique may be used in humans.

There are a number of technical problems that must be solved before the full promise of gene therapy is realized. Many genes only function in a specific tissue, and it may be difficult to insert "repair" genes into these tissues. The central nervous system, in particular, may be difficult to target because it has a "blood/brain barrier" that prevents the genes from being inserted. In addition, if the gene therapy is to be permanently effective, the repair gene must be inserted into a cell that divides to produce a colony of additional cells that will express the gene. Gene therapy for cystic fibrosis presents this type of difficulty. It may be possible to insert the repair gene into the surface cells of the affected individual's airways, but the function of the gene will be lost as these cells are shed. In order to effect a permanent cure, the cells that divide to produce the lining of the airway itself must be genetically altered. These cells have not yet been found and, once found, are likely to be less accessible for gene insertion than are surface cells. Gene therapy for cystic fibrosis continues to involve repeated treatments for an indefinite period of time—perhaps every month or two for life.

Another obstacle to effective gene therapy is that many genes function in a complex system that requires feedback from other parts of the body. Unless inserted genes are specifically targeted to locations in the genome, or unless genes are inserted with appropriate regulatory sequences, the inserted gene may not function properly. Simply inserting genes for the production of insulin, for example, would not benefit diabetics unless there were also mechanisms to carefully coordinate the insulin production with the individual's blood sugar level.

Finally, random insertion of genetic material into the genome, which is the current scattershot method, may disrupt other genes. This may result in the creation of a new health problem such as cancer secondary to the disruption of a gene responsible for cell growth regulation.

While these problems are formidable, there is no reason to believe they cannot be solved in the next ten to twenty years, at least for some genetic conditions. For example, research to enable genes to be inserted into a desired site in the genome is making good progress.[10] A variety of single gene defects may be amenable to gene therapy in the near future.

We have been discussing gene therapy primarily in the context of altering "somatic" cells, that is, cells in specific tissues of the body. A different type of gene therapy that is much more controversial is "germ cell" therapy.[11] With this approach, reproductive cells, that is, eggs or sperm, would be altered genetically to eliminate disease-producing genes. This could be done in two ways. First, the eggs or sperm of an individual with a genetic condition could be directly targeted by gene therapy. In the individual with cystic fibrosis, for example, the normal gene would be inserted into the lungs, but also into sperm-producing cells in the testes or into the eggs in the ovary. The objective would be to prevent the cystic fibrosis gene from being passed on to the next generation.

An alternative way to insert genes into germ cells is to do so very early in the development of the embryo. If a gene is inserted into the cell of an embryo, then all the cells that derive from that early cell will contain the new gene. If the cells that derive from the treated embryonic cell produce the reproductive system, then germ cell alteration would have occurred—the eggs or sperm of the resulting individual will be free of the abnormal gene.

The notion of germ cell gene therapy has been controversial. There is a deep concern about altering the genetic makeup of future generations. Some believe that this power treads too close to the powers of God or Nature. Others are concerned about the potential harms from manipulating the germ line. Genetic interactions are highly complex and scientists have only a limited ability to perform the necessary technical feats. Mistakes will be made and the mistakes may be passed on to future generations. The discussion of these concerns is beyond the scope of this book. However, there are other significant practical limitations to the application of germ cell gene therapy that are relevant here.

First, it is useful to look at the purposes for germ line gene therapy. One stated purpose of germ cell alteration is to prevent transmission of the genetic abnormality to the next generation. However, for autosomal *recessive* conditions like cystic fibrosis and many other hereditary conditions, gene *carriers* as well as individuals affected by the disease would need to undergo germ line alteration. This would be a massive undertaking. The same results could be obtained through widespread prenatal screening and selective abortion, or through *in vitro* fertilization for at risk couples and selective implantation of unaffected embryos—the technique used by the Brittons in Chapter One.

The most appropriate candidate for germ line therapy for recessive conditions would be the rare couple in which *both* partners were affected. All of the children of such a couple would be affected with the disease, so germ line therapy might be attractive to produce an unaffected child.

Autosomal *dominant* conditions, like Huntington's disease, offer a somewhat better opportunity for decreasing disease in the next generation through germ line alteration. If the germ cells of an affected individual were genetically altered, then that individual would not be at risk of passing the gene to the next generation. Again, however, the same results could be obtained in other ways—prenatal screening and selective abortion, IVF and selective implantation, or through somatic cell treatment of affected children. It is by no means obvious that germ line treatment would have clear advantages to these alternatives.

In addition to the marginal indications for germ line gene therapy, there are substantial technical barriers to overcome. The normal male produces sperm from billions of sperm stem cells, although only one sperm ultimately fertilizes the egg. In order to assure that this one sperm does not contain the abnormal gene, a very high percentage of the sperm stem cells would need to be altered without disrupting their normal functions. At the present time, gene insertion can be achieved only in an extremely small percentage of target cells. For a male to be confident that germ line therapy would work, drastic improvements in the rate of gene insertion into sperm stem cells would be necessary. This is unlikely in the foreseeable future unless there is a great demand for this technology.

The alteration of eggs is even more problematic. A woman's eggs complete most of their development before the woman is even born. The eggs remain in a largely dormant state until near the time of ovulation. Gene insertion techniques typically rely on the manipulation of DNA while it is actively dividing. The accurate and reliable insertion of genes into dormant egg cells presents a formidable technical challenge. Again, in the absence of a great demand for this type of intervention, it seems unlikely that the benefits of this technology development will seem worth the risks.

Germ line therapy, achieved through the insertion of genes into affected embryos, also may not prove worthwhile. The process of *in vitro* fertilization generally produces several eggs. If prospective parents, reproducing by means of *in vitro* fertilization, produced an embryo that was affected with a genetic condition, the parents would likely choose to discard that embryo, rather than attempt to genetically manipulate the embryo, implant it, and hope that no ill effects were produced for the resulting child. The most likely potential use of germ line therapy would be in an embryo that was conceived in the conventional manner and found, after testing, to be affected with a genetically related disorder. If the parents of such an embryo did not want the disorder to be passed on to succeeding generations, and yet were unwilling to abort the embryo, germ line therapy would be attractive. But demand of this type is not likely to be great.

While there appear to be few medical reasons to employ germ line gene therapy for genetic diseases, germ line applications potentially would make more sense for genetic *enhancements*, that is, genetic alterations that improve what were already normal traits. Since genes work in concert with one another, genetic enhancements may be more effective for many traits if the genetic material is inserted early in the development of the embryo. This would result in germ line alteration as well. This brings us to the general topic of genetic enhancement.

GENETIC ENHANCEMENT

Genetic enhancement may offer the most tantalizing possibilities for shaping our genetic landscape. Once relegated to the province of science fiction, and although not yet part of the present, genetic enhancement will likely be applied to human embryos within the next twenty to thirty years.

What is genetic enhancement? Earlier in this chapter, we described efforts to introduce healthy genes in those having an abnormal gene, in an attempt to correct a disease condition. Genetic enhancement involves no abnormal gene and no disease condition. Instead, its goal is to amplify "normal" genes in order to make them "better." For example, suppose that a man having a normal gene for height grows to five feet, ten inches. This male would be deemed to be within the normal range of height for his gender, "normal" being a somewhat fluid concept and encompassing a range of values. However, if he wished to become a basketball player, his height, which for other purposes is "normal," would likely be a barrier. If, through genetic manipulation, this male could attain a height of six feet, six inches, the process could not be seen as having corrected a genetic defect, for his gene was not defective. Instead, the process would have amplified a normal trait. This is genetic enhancement.

As yet, we do not know how genetic factors contribute to specific characteristics, such as shyness or perfect pitch. Many of our physical and psychological traits are a complex mix of genetic and environmental factors. We can only imagine the ways in which genetic enhancements might change who we are and who our progeny will be. With the addition, subtraction, or modulation of specific genes, we might be able to influence such characteristics as longevity, endurance, wound healing, intelligence, personality, and eye color.

Weight control is an area in which the potential for genetic enhancement is accelerating. There is great interest in obesity research from a health standpoint. Obesity is a serious health problem in developed countries—1 in 3 Americans is obese.[12] Another large segment of the population is overweight, but not sufficiently so to be called obese. Obesity increases the risk of heart disease, diabetes, and orthopedic problems. Obesity extracts a high toll in the expenditure of health care dollars, in an effort to address the conditions caused by or exacerbated by obesity.

In 1994 scientists identified a gene in mice, the *ob* gene, that appears to have a significant influence on obesity.[13] Mice who inherited a mutated copy of this gene grew up to three times fatter than did mice having a normal *ob* gene.

The *ob* gene appears to code for a protein secreted in fat tissue. This protein may function by binding to centers in the brain responsible for appetite control. Investigators hypothesized that as the body creates more fat tissue, protein secretion increases, which in turn decreases food intake by regulating the appetite control center. Mice with mutated copies of the gene are unable to create this protein. As a result, their appetites stay "turned on" despite large quantities of fat tissue. When obese mice were injected with the *ob* protein, they experienced a 30-percent weight loss. Further, when the protein was injected into mice with normal *ob* genes, they too lost weight. [14]

Humans have a gene that is almost identical to the *ob* gene, although its exact function is as yet unknown. Consistent with the findings of an "obesity gene," one's body size is significantly influenced by genetic factors. Thin parents tend to have thin children, and overweight parents tend to have overweight children. This could reflect social and environmental factors, but research has shown that the weight of adopted children correlates more closely with their biologic parents than with their adoptive parents.

How then might our bodies be genetically altered to reduce obesity? At present, scientists have not yet ascertained the precise mechanisms involved in obesity, but there are a number of possible methods by which the gene could be affecting obesity. The *ob* gene may affect the efficiency of the protein's structure, or may control regulatory sequences that turn on or shut off protein production. Perhaps the *ob* gene influences the brain's appetite control receptor molecules. Once points in the appetite control chain that control obesity are identified, the ability to manipulate the elements can be devised. One possible approach to genetic manipulation would be to insert genes capable of highly efficient production of the *ob* protein into a human embryo early in development. The genetic insertion would be targeted to the genome in a manner to disrupt and replace the embryo's inherited copies of the *ob* gene. With a brisk production of the *ob* protein, the resulting individual would be much less likely to be obese.

A second possible approach would be to insert the *ob* gene into an implantable tissue reservoir, perhaps into the lining of an artificial blood vessel, as described earlier in this chapter. The *ob* gene could be designed to function continuously to suppress appetite, but it also could have linked regulatory genes inserted that could increase or decrease *ob* protein secretion by drug administration, if necessary.

Implantation of engineered tissues would significantly decrease the complexity of the genetic enhancement in comparison to embryo manipulation and presumably would make it safer by making it reversible: that is, if the effect were no longer desired, the tissues could be removed.

Individuals with the abnormal condition of obesity, of course, would not be the only persons interested in this gene. Overweight individuals, those whose condition is not medically serious enough to classify them as obese, would likely also choose to correct this state. Even those whose weight was normal, but who desired to be thinner for professional or cosmetic reasons, might seek *ob* gene engineering. As we move on a continuum from the obese and the overweight to those who fall within a normal range for weight but who wish to move closer to a more desirable end of the normal distribution, or to those who only achieve a desirable weight through tedious exercise and/or the deprivations of dieting, we move from treatment for genetic disease to treatment for genetic enhancement.

For biotechnology and pharmaceutical companies, the potential rewards from this research are enormous. Our society spends an enormous amount of money on self-enhancement efforts, including diet, exercise, and plastic surgery. Weight loss aids and programs are multimillion-dollar businesses. If a pharmaceutical solution is developed, the demand will be great, as will the opportunities for great profits for those who offer it for sale. The results of the *ob* gene research already have prompted a biotechnology company, Amgen Inc., to pay $20 million to Rockefeller University for an exclusive license to develop products derived from this gene.[15]

The same incentives exist for other enhancements as well. Currently, drugs such as steroids, Prozac, and a variety of purported "intelligence enhancing" agents are being used to give people a boost—an edge in an increasingly competitive world. As more is learned about personality traits and intelligence, genetic enhancement of these traits may become possible.

CONCLUSIONS

The full flowering of these technologies will not occur for some years. Numerous improvements in genetic engineering will be necessary before we can safely alter the genes of human embryos. Even after the

Human Genome Project is completed and all the genes are mapped and sequenced, it will take years of painstaking work to understand the interaction of multiple genes in producing complex, polygenic conditions and traits.

Two facts are striking, however. One is how quickly progress is being made. The Human Genome Project is expected to be completed by 2005, on time, if not ahead of schedule.[16] The press regularly reports the discovery of new genes associated with widespread and often catastrophic disorders, such as breast and colon cancers and Alzheimer's disease.[17] Experiments using gene therapy are underway across the country and around the world.[18] Even scientific skeptics, who caution that we should not expect to see techniques such as genetic enhancement available anytime soon, admit that these technologies may be developed well within our lifetimes.[19]

The second striking fact is how many of these technologies are available now. A number of genetic tests are sold commercially for disorders such as Huntington's disease, cystic fibrosis, Down syndrome, Tay–Sachs, sickle cell disease, fragile X, polycystic kidney disease, muscular dystrophy, and hemophilia.[20] Drugs that used to be rare and prohibitively expensive are being mass-produced using recombinant DNA technology.[21] While direct manipulation of genes to enhance desirable human traits has not yet been attempted, there have been a number of reports of parents seeking human growth hormone, produced through recombinant DNA, to increase the height of children whose height already is within normal range.[22]

So far, we have discussed new and potential genetic techniques and their effects on an individual's health and well being. These techniques also will have a major impact—both good and bad—on society. It is to this that we now turn.

4
The Impact of Genetic Technologies

The previous chapter described the new technologies that can be expected to flow from the genetic revolution. These technologies will create unprecedented benefits and risks. Critics of the genetic revolution warn of terrible dangers that could arise if mistakes are made in developing and perfecting these technologies. New and terrible genetic scourges may be unleashed on mankind. We may create monsters. We may damage our genes so that the we cannot survive future assaults of disease. We may homogenize the human gene pool so that the species lacks the diversity to respond to future environmental challenges.[1]

In a realm of scientific endeavor as virgin as decoding and manipulating the human genome, we must take these concerns extremely seriously. The utmost care is required in designing, carrying out, and monitoring the results of genetic research. We must be on guard to protect the rights of experimental subjects, especially when human genetic experiments are carried out on children, embryos, fetuses, and others who cannot protect themselves. We must have a clear understanding of the risks and benefits of genetic research and of the interventions it produces, so that we may avoid doing more harm than good.

At the same time, we cannot be blind to the fact that genetic technologies offer unprecedented opportunities. Individuals with access to information about their genetic endowments will be able to gaze into the future and predict their susceptibility to genetically related disorders. Couples will be able to learn the genetic characteristics of their fetuses and selectively abort those with dread genetic diseases. Those couples fortunate enough to have access to *in vitro* fertilization will be able to choose among embryos and implant into the mother's uterus the one that is free from significant genetic impairment. Carrier testing, such as that now being carried out on Americans of European Jewish descent, can prevent the conception of children with genetic disorders such as Tay–Sachs. Therapeutic advances in genetics will permit the treatment and ultimately the prevention of numerous diseases. Identifying genes associated with nondisease characteristics will make it possible to design pharmacological and perhaps even genetic interventions to enhance positive traits and eliminate or reduce those that are undesirable.

Barring a catastrophe, such as the unintentional widespread release into the environment of a lethal, new virus, or the discovery that unscrupulous researchers are conducting unethical human genetic experiments, the potential benefits from these technologies will generate overwhelming pressure to carry on the work despite the risks. If this research meets with any substantial degree of success—and there is no known scientific reason why it should not—we can expect to see the development of an ever-increasing array of effective, new genetic technologies.

Yet these very accomplishments will lead to new challenges. The same unprecedented power that makes genetic technologies so alluring also makes them dangerous. The threat in this case is not the risk of research errors and experimental disasters. On the contrary, it is the very success of the research, the triumph of our scientific efforts. Indeed, the greater our success in developing genetic technologies, the greater these threats will become.

These threats revolve around the central question of whether or not an individual has access to genetic technologies. One type of threat occurs when individuals are given genetic technologies that are contrary to their interests. A second type of threat occurs when individuals are denied access to technologies that would benefit them. Both of these threats create the same risk: that individual fates will vary in profoundly important and potentially negative ways, depending on the individual's access to genetic technologies. As we will explain in the following section, the precise nature of these threats varies according to the specific genetic technology in question.

GENETIC TESTING AND SCREENING

One of the first practical accomplishments of the Human Genome Project has been an increase in the ability to determine whether an individual is suffering from a genetic disease, and whether the individual's DNA contains genetic material that predisposes that individual, or her offspring, to certain genetic diseases and disorders. As the Human Genome Project unfolds, the number of conditions for which a test exists is increasing, and the accuracy of the tests is steadily

TABLE 1.   Genetic Disorders For Which There Are Tests

Disease	Incidence	Ethnic Predilection
Alpha–1 Antitrypsin	1/2,500–5,000	Caucasians
Caravan Disease	1/5,000	Ashkenazi Jews
Cystic Fibrosis	1/2,500	Caucasians
Duchenne Muscular Dystrophy	1/3,500 males	None
Fragile X	1/1,250 males 1/2,500 females	None
Gaucher Disease	1/450	Ashkenazi Jews
Hemophilia A	1/5–10,000 males	None
Huntington's Disease	1/10,000	None
Myotonic Dystrophy	1/8,000	None
Neurofibromatosis I	1/3–5,000	None
Sickle Cell Disease	1/600	African Americans
Spinal Muscular Atrophy	1/6,000	None
Tay–Sachs Disease	1/3,600	Ashkenazi Jews

improving. Ultimately, we should be able to identify and accurately test for all genetically related maladies.

A great deal of the work to date on the ethical, legal, and social implications of the Human Genome Project has focused on genetic testing. This has stemmed in part from the fact that the ability to detect some genetic abnormalities predated the start of the Human Genome Project by a couple of decades. This has given us more time to become aware of and to grapple with the ethical, legal, and social issues that are raised. Prenatal testing for Down syndrome, for example, has been available since the 1960s.[2] In the early 1970s, tests were developed to identify carriers of sickle cell disease.[3] A list of common genetic disorders for which tests are available is presented in Table 1.[4]

The public attention that has been given to genetic testing, by and large, has emphasized the negative. Genetic information can be complex and difficult to understand, leading individuals to misunderstand test results. For example, they may assume that a positive result invariably means that they will suffer from the disorder, when, in fact, it may signify merely that there is a certain probability that they will display symptoms of the disorder in the future. When sickle cell testing became available in the early 1970s, many people mistakenly thought that a positive result meant that the individual in question had sickle cell disease, when it merely established that the individual was a carrier of the disease trait and could pass the aberrant gene on to his or her children.[5] Even when a positive test result for a condition such as Huntington's disease indicates that the individual will suffer from the symptoms of a genetic disorder in the future, the test seldom reveals how seriously the individual will be affected or when the disease will manifest itself. To guard against these mistakes, health professionals uniformly have urged that genetic testing be accompanied by careful counseling, both before the test is administered and after the results are available.

The predictive power of genetic tests also makes their results useful to third parties. Statistical data on the prevalence of genetic disorders in the population can assist the government in projecting the need for treatment facilities and personnel in the future. Public health officials have used similar data on the spread of HIV infection and AIDS to estimate what resources they will need to care for the affected population.

But third-party interest in genetic information is not always so innocuous. Insurers may require applicants for insurance to be tested to determine their susceptibility to genetic disorders. Positive results can lead health insurers to refuse to insure the individual, to charge prohibitively high premiums, or to regard the individual's susceptibility as a preexisting condition which they will not cover or will cover only to a limited degree. Positive test results also can preclude the individual from obtaining life insurance, or at least from being able to purchase a policy at an affordable price.

Employers have interests in genetic information similar to those of health insurers. Employers may refuse to hire persons with genetic predispositions to disease in order to avoid having to pay the costs of future treatment, either through higher insurance premiums or in the form of direct payments by employer self-insured health plans. For the same reason, a positive genetic test result may cause an employer to try to fire a current employee. Employers also may fear that an affected employee may create a safety risk for customers and other employees, if the disorder is one that, like Huntington's disease, can impair coordination and judgment.

Federal law prohibits some types of discrimination by employers based on the results of genetic testing. The Americans with Disabilities Act (the ADA), for example, makes it illegal for an employer to refuse to hire or to fire someone on account of a genetic disability, unless the disability makes it impossible for the individual to perform the job. It is not clear, however, whether legal protection extends to employees who have tested positive but who have not yet exhibited symptoms of a genetic disorder and may never do so or may not do so for years. Initially, the federal agency charged with interpreting and enforcing the ADA—the Equal Employment Opportunity Commission—took the position that these "presymptomatic" individuals were not protected.[6] Recently, the EEOC has indicated that it has changed its mind and warned that employers may not discriminate against presymptomatic as well as symptomatic individuals.[7] As yet, there have been no enforcement actions by the EEOC, so the courts have not had a chance to determine if the EEOC's interpretation is consistent with the law.

Even if the law protects individuals with positive genetic test results from outright employment discrimination, they may have little recourse against health insurers who deny health care coverage. The ADA contains an exception for insurance. The law permits insurers (including employer self-insured plans) to take disability into account in their insurance underwriting.[8] As a result, insurers may refuse to extend coverage to persons with genetic disabilities or susceptibilities because it would increase the insurers' costs. On the other hand, the law goes on to prohibit insurers from using their underwriting activities as a "subterfuge" to get around the intent of the ADA.[9] This provision has yet to be interpreted by the courts, and, therefore, no one is certain yet what it means. At the least, it seems to prohibit insurers from refusing to extend coverage to persons with disabilities at the same time that insurers cover nondisabled persons with equally expensive medical conditions. The EEOC has gone farther, suggesting that the subterfuge clause may forbid the insurer from denying coverage unless, by raising costs or forcing the insurer to increase its premiums, this would significantly jeopardize the insurer's ability to stay in business.[10]

Those in charge of the Human Genome Project have been well aware of the threat of insurance discrimination created by genetic testing. In 1993 the National Institutes of Health convened a task force to explore the problem and make policy recommendations. The task force concluded that the problem was potentially serious and recommended that laws be passed to prohibit insurers from requiring individuals to undergo genetic testing, or from basing coverage decisions on the results of such tests.[11] Several states already have these types of laws on the books.[12] Similar bills have been introduced in Congress. But insurers vigorously oppose these efforts, arguing that they should be allowed to offer insurance at cheaper prices to persons who are not susceptible to genetic disorders, and that the only way to do this is to enable insurers to obtain and act upon the results of genetic tests. In some of the states that have enacted statutes prohibiting insurers from discriminating on the basis of genetic testing, insurers are allowed to base their insurance decisions on a person's family history, which is merely another, albeit more crude, source of genetic information.

Employers and insurers are not the only parties with an interest in the results of genetic tests. Family members may seek another member's test results in order to learn if they themselves are at risk. This information can be beneficial to the family, by leading potentially affected individuals to get tested themselves and to take preventive measures if the results are positive. Genetic information also can disrupt family relationships. Children may resent parents who have passed on to them inherited susceptibilities to illness. Marriages may break apart when one person finds out that a spouse is a carrier or is presymptomatic for a dread genetic disease. Prospective mates may insist on genetic testing before they will agree to enter into a marriage relationship in the first place. Genetic information can profoundly disrupt familial relationships when a genetic test undertaken to detect a person's risk for a genetic disorder happens to reveal that the person's father is not the man previously identified as such.

A more frightening prospect is that genetic information may be misused by the government. The state has a number of reasons to seek access to information about an individual's genetic makeup. As mentioned earlier, public health officials can use this information to plan for future health care needs. At first, this may seem to pose no threat to the interests of individuals being tested; public health officials are only interested in population statistics, so genetic data can be collected and maintained anonymously. But in order to be compiled accurately, public health data must be collected and stored identifiably—that is, linked to the names and addresses of the persons who were tested. Otherwise, there is no way to prevent the same individual's test results from being counted more than once. For this reason, some public health officials have opposed anonymous testing for HIV, where the person being tested does not reveal his name and is known only by a code number.[13] Information about HIV tests is confidential and is supposed to be released only in an aggregate manner so that individual identities are kept secret, but the fact that the information is collected and stored in an identifiable fashion means that there is always a risk that individual identities someday could be made public, or that they could be made available to third parties without the individual's consent.

The state may have other uses for genetic information beyond compiling it for statistical purposes. For example, the government may require persons to be tested for disorders that can be treated if detected early enough. Such an approach would save money by avoiding the need for more expensive health care later on. This type of testing program is already in operation. As noted in Chapter Two, state laws require that newborns be screened for a variety of treatable genetic disorders, such as PKU and hypothyroidism. As the Human Genome Project produces larger and larger numbers of genetic tests for treatable diseases, state governments also may make these tests mandatory.

Treatment does not necessarily have to await birth. Increasingly, we are developing means of treating fetuses *in utero*. For example, fetuses have been treated successfully with prenatal transfusions for anemia from maternal/fetal blood incompatibilities. In addition, fetuses have had surgical procedures *in utero* for blocked kidneys, diaphragmatic hernias, and hydrocephalus with variable success. These genetic disorders can be detected with ultrasound. Currently, these tests are only administered at the request of the parents. But as the ability increases to treat fetuses more effectively or more cheaply for genetic ailments, the government may require prenatal genetic testing by law.

Prenatal or newborn genetic testing need not be limited to physical disorders. As we learn more about the human genome, we almost certainly will identify genetic tests to detect treatable psychological disorders. The government could require testing for these disorders, particularly if they cause affected persons to engage in violent or criminal behavior.

Many experts on medical ethics believe that the government should not mandate genetic testing for fetuses or newborns. In 1993 the prestigious Institute of Medicine of the National Academy of Sciences recommended that, at most, the government should mandate that people be *offered* genetic testing for newborns, and then only for "disorders with treatments of demonstrated efficacy where very early intervention is essential to improve health outcomes."[14] The report argued that there is no need for the government to require that parents test their newborns since they have an interest in obtaining effective treatment and will obtain testing voluntarily so long as they are made aware that testing is available.

Two emerging themes in health care policy may lead the government to mandate that testing be offered not only for treatable diseases but for nontreatable diseases as well. One is the need to reduce health care spending. The second is the growth of the so-called "communitarian" movement, whose adherents, including several prominent medical ethicists, believe that individual liberty often must give way to the needs of the community.[15] These views might lead the government to require that people be tested to see if they are carriers for genetic disorders even though the disorders could not be treated. The objective would be to discourage such people from having children, or at least from having children with other carriers. This would reduce the number of births of genetically afflicted children, thereby saving money as well as arguably reducing suffering. A policy of discouraging carriers from having children also would gradually eliminate the genetic mutation in question from the population.

There is historical precedent for this type of government program. In the early 1970s, several states passed laws requiring African-Americans to be screened for sickle cell trait, despite the fact that there was no effective treatment for the disease.[16] The only real benefit from the screening was to identify people who were carriers for the disease in order to discourage them from having children with other carriers, since their children would have a 25 percent chance of being afflicted with the disease and a 50 percent chance of being carriers themselves. Linus Pauling, the Nobel laureate, even advocated that carriers of the sickle cell trait be tattooed with an *S* on their foreheads so that they would be able to identify each other and avoid procreating.[17]

The mandatory sickle cell screening programs were widely criticized. Many people incorrectly thought that a positive test result invariably meant that they were afflicted with the disease, when, in fact, they might only be carriers of the trait. Concerns also were raised about the threat of discrimination against those who were carriers.

It is easy to see how these negative uses of genetic testing, both private and public, could undermine fundamental notions of social equality. Discrimination on the basis of genetic endowment means that persons with positive test results would be denied benefits that were enjoyed by those who tested negatively, or perhaps were not tested at all. Society risks becoming divided into those who are genetically sound and those who are genetically afflicted. The former would have an easier time getting jobs and insurance. The latter might be scorned because they had nontreatable disorders and therefore were doomed by their genes to suffer, or because, although their disorders were treatable, they were genetically inferior because they needed treatment in order to be healthy, or because they were carriers who chose to impose the costs of future genetic abnormalities on society by not abstaining from reproduction.

So far we have emphasized the negative potential of genetic information—the risk of abuse by public and private interests stemming from the growing predictive power of genetic testing. But we must not overlook the tremendous benefits that genetic testing potentially can provide. Even if the genetic disorder cannot be treated, knowledge of one's risk can enable an individual to make important decisions in areas of family planning, reproduction, financial planning, and life-style choices. If someone finds out that she is a carrier for a serious genetic disorder, such as Tay–Sachs, she may insist that a potential mate be tested and refrain from procreating with that person if the test is positive. A person who discovers that he or she bears the gene for Huntington's disease can take steps before the symptoms manifest themselves to provide for her family and her eventual incapacity, to make advance decisions about medical treatment, to eschew career opportunities that provide the bulk of their rewards later in life, and so on. A couple undergoing *in vitro* fertilization can decide not to implant embryos with known genetic abnormalities, while a couple that knows that a fetus is afflicted with a dread disorder can choose to have an abortion. None of these options is possible without some degree of genetic information.

Genetic information also is important as a diagnostic tool. Children diagnosed with developmental delay or failure to thrive may have a variety of genetic tests included in their evaluation. Identifying a specific etiology for such problems can assist in treating, preventing, or ameliorating future problems, and in informing the family about reproductive risks for subsequent children.

Even more significant is the fact that genetic testing may reveal the need for treatments to prevent or mitigate the severity of a genetic disorder in a presymptomatic individual. A person who tests positive for the BRCA1 gene for breast cancer, for example, might be advised to increase dramatically the frequency of breast examinations and mammograms, and might even decide to undergo a prophylactic mastectomy—that is, to remove her breasts before cancerous tissue is spotted. Genetic testing in this sense would be performing a "gatekeeper" function: persons would need the testing to know that they needed further interventions.

If the potential negative uses of genetic information create the threat that some people will be forced to undergo genetic testing against their will, or that test results will be used to discriminate against them, the positive uses of genetic information raise the concern that some people might be denied genetic testing that they desire. The high cost of testing, for example, might place it outside the reach of those who lack generous health insurance plans or who cannot afford to pay for the testing themselves. Physicians and other health professionals who are opposed to abortion might withhold information about the availability of genetic testing because they fear that the test results might lead a couple to abort an affected fetus. In a somewhat similar scenario, some health professionals may refuse to provide prospective parents with information about the gender of their fetus to prevent the abortion of an unwanted boy or, more often, girl.[18] Even if health professionals were willing to inform people about genetic testing, people who lacked access to health care may be unaware of the availability of genetic testing. Those who obtained genetic testing may misunderstand its results. In short, the benefits of genetic information are likely to be available only to some people: those who know about genetic testing, who know how and where to get it, who can afford it, and who can assimilate the results accurately.

As discussed before, genetic information used for negative purposes threatens to divide society into the genetically sound and the genetically afflicted. Similarly, if only some individuals obtain the benefits of genetic information, society will be divided into those who have access to desired testing and those who do not. The latter might still be tested, but the testing would be involuntary or for negative uses. Although it is possible that these persons might derive some benefit from the results of unwanted testing (for example, they could use the test results to make reproductive decisions), they would be denied the option of choosing to be tested for positive purposes.

GENE THERAPY

As Chapter Two described, the Human Genome Project gradually is yielding advances in genetic medicine beyond simply the ability to detect genetic abnormalities. Once the gene is identified, it becomes possible to identify the protein for which it codes. The next step is to produce the protein through techniques such as recombinant DNA manufacturing and to introduce the protein into the systems of individuals who suffer from disorders caused by its lack. A more sophisticated technology is gene therapy, which involves introducing healthy genes into the body, so that the genes themselves stimulate the production of the protein. In this way, it may be possible eventually to cure many genetic disorders.

Like genetic testing, gene therapy has both negative and positive potential. In order to reduce health care costs, the government might require individuals to undergo gene therapy without their consent, particularly if they were insured through public health care programs such as Medicare, Medicaid, or CHAMPUS. The most likely targets of government-mandated treatment for genetic disorders would be newborns and fetuses. They are incapable of giving consent to treatment, and the government could justify intervening without parental consent on the basis that parental refusal would be contrary to the best interests of the child. There is already some precedent for such an approach. State laws require eyedrops to be placed in the eyes of newborns to protect them against gonorrhea and chlamydia.

Mandatory treatment of newborns and fetuses might not seem contrary to their interests. However, it poses a number of risks. Gene therapy would be preceded by genetic testing to reveal the anomaly to be treated and could lead to the same types of discrimination that might result from genetic testing alone, particularly if the therapy could not effect a complete cure. Forced treatment would interfere with parental decision-making authority. While this may seem appropriate when the treatment is clearly beneficial and the disorder to be treated is serious or life-threatening, it may not be as justifiable when the treatment itself creates significant health risks or when there is substantial disagreement within the medical community over the value of the treatment. Finally, and perhaps most importantly, forced treatment of newborns and fetuses sets a precedent that could be extended to those persons capable of giving their consent, or to interventions that were not so clearly therapeutic, such as those designed to alter an individual's personality.

The government is less likely to require adults to be treated against their will. The only precedents in this country are laws that permit government officials to quarantine and forcibly treat persons with communicable diseases, such as syphilis and tuberculosis, in order to protect the public health. Genetic disorders are not ordinarily regarded as being communicable. You cannot "catch" a genetic illness by being near or having contact with someone who is afflicted. Therefore, there would seem to be little motivation for public health officials to protect the public by forcibly treating affected individuals. However, genetic disorders can be transmitted via reproduction from one generation to the next. Thus, they are said to be "vertically," but not "horizontally," transmissible. The desire to contain the vertical spread of genetic disorders might lead government officials to require infected persons to be treated by reducing or eliminating their ability to reproduce, by requiring them to undergo selective abortion of affected fetuses, or by forcing them to have their genetic anomalies corrected in their reproductive cells.

Gene therapy also creates the risk that risky experiments will be performed on people who cannot protect themselves, such as those who are poor or imprisoned, or who, because they lack information or have impaired mental faculties, are incapable of giving their informed consent to participation. There are numerous historical examples of such experiments, ranging from the infamous Tuskeegee syphilis trials, begun in the 1930s, in which men were denied treatment in order to chart the course of the disease, to LSD and nuclear radiation studies on unwitting servicemen.[19] Human gene therapy is likely to require a substantial amount of trial and error before its risks can be reduced to levels that justify making it widely available, and the burden of the mistakes may fall disproportionately on those who are unable to protect themselves.

Despite all of these negative concerns, gene therapy has the obvious potential to be enormously beneficial to those who receive it.

Persons who obtain access to gene therapies can avoid or mitigate the effects of disease in ways at least as dramatic as the first administrations of penicillin in the 1930s.[20] Access to gene therapy would be the key to good health and even to survival. Even when still in their experimental phase, genetic treatments might be the only hope for those afflicted with serious or life-threatening illnesses, many of whom might be eager to accept known or unknown risks in return for potential benefits. Indeed, instead of the conventional fear that the only people who would receive these treatments would be the poor and other relatively powerless populations who could not protect themselves against questionable experiments, we might be more concerned that those who are wealthy and powerful would reserve for themselves the privilege of obtaining access to the latest experimental treatments.

Society, therefore, would be divided into those who had access to gene therapy and those who did not. The former not only would be healthier, but they might be regarded as being healthier in a qualitatively different way. By correcting the defective genes themselves, gene therapy would cleanse those who obtained access to it of genetic defects. They could be said not only to be healthier, but to be genetically healthier.

Lack of access to gene therapy could give rise to the same types of discrimination that would result from genetic testing—social stigma, loss or lack of jobs and insurance, and so on. But there is a crucial difference: in the case of genetic testing, the persons who would face discrimination would be those who had received testing for genetic disorders and who had tested positive. In the case of gene therapy, however, those who suffered discrimination would be those who had *not* received gene therapy.

GENETIC ENHANCEMENT

The final major category of genetic technologies is genetic enhancement—the ability to use genetic techniques to improve genetically related traits that, without improvement, would lie within normal ranges. If the traits that could be enhanced were expected to yield significant social benefits, there is a chance that the government would require people to be enhanced against their will. These people arguably might be said to be harmed by receiving enhancements that they did not desire, on the grounds that the government deprived them of control over their own lives. But like the threat to notions of social equality posed by gene therapy, the real threat lies not in receiving enhancement, but in *not* receiving it. Those who were denied enhancement would be disadvantaged compared to those who were enhanced. Depending on what traits were amenable to enhancement and how effective the enhancement was, the advantages enjoyed by those who were enhanced could be so significant that they were determinative of social status and success.

The development of enhancement technologies will lag behind the search for disease-related genetic tests and treatments. The government is bound to devote the bulk of its substantial research funds toward eradicating genetic disorders rather than creating enhancement technologies. Many genetically related traits that would be considered candidates for enhancement are likely to be multifactorial—that is, the product of interactions among multiple genes. This makes it difficult to identify the responsible genes and to manipulate the associated traits.

The threat to social equality represented by genetic enhancements, although not imminent, cannot be dismissed as impossibly far in the future, however. Geneticists actively are exploring multifactorial genetic disorders. The knowledge that they are acquiring about how genes interact, and how their disease-producing interactions can be identified and manipulated, will be available to accelerate the search for genetic enhancements as well. Even if the government concentrates its research grants on genetic disorders, private industry is bound to invest heavily in enhancement research in view of the potential commercial rewards.

Support for these predictions can be found in the practices surrounding the use of genetically engineered human growth hormone (HGH). Originally, this substance was extremely scarce, since it had to be extracted from the pituitary glands of cadavers, and this could only be done legally in conjunction with an autopsy. However, since 1985 we have been able to make bacteria produce HGH using recombinant DNA techniques, making it relatively abundant.

Now that HGH no longer is in such short supply, some researchers and clinicians have shifted from offering it only for children with HGH deficiencies, such as pituitary dwarfism, to providing it to the parents of children who are outside the normal range of height for that age.[21] Genetic scientists and bioethicists are even contemplating its use in normal children to make them taller, thereby improving their ability to play sports such as basketball.[22] The authors have seen television commercials urging parents to call a physician's 800 number to find out how HGH can increase their children's stature.

HGH may be useful for other purposes besides increasing height. New research suggests that HGH and other hormones may slow the aging process.[23] Much of the interest is aimed at improving the strength and stamina of persons who are aging normally and not just those who are especially frail or who might take HGH to accelerate their recovery from illnesses or injuries.

Genetic science has a long way to go before genetic enhancements are developed, are proven safe and effective, and become widely available commercially. The use of HGH is not without health risks. There are side effects, some serious, although these seem to be rare. Little is known about long-term effects in children who are not HGH-deficient. Furthermore, a modest degree of enhancement, such as extra height for individuals already within normal limits, confers only limited social advantages. But research on genetic enhancements is destined to accelerate. It is likely to be only a matter of time before researchers discover methods for enhancing traits with much more clear-cut personal and social potential. The allure of a genetically engineered substance that slows and perhaps even reverses aging, for example, is undeniable. So are its commercial possibilities.

But what if genetic enhancement were readily available only to certain individuals? Enhancement would give its possessors advantages over the unenhanced. Depending on the nature of the traits that could be enhanced, these individuals could be stronger, handsomer, smarter, more graceful, better able to withstand stress, less prone to fatigue. They could have better eyesight, hearing, and memory.

Enhanced traits arguably would improve the possessor's quality of life. A person who had more strength or stamina might be more active, play recreational sports more aggressively, or be the life of an all-night party. More importantly, however, individuals with access to effective genetic enhancements would be likely to fare better in competing for scarce societal resources. They would be more qualified for jobs requiring enhanceable skills. Their enhanced physical attributes would make them more attractive to a greater number of potential mates. They would be able to amass wealth more easily, either by earning it or by marrying into it.

The social impact of these genetic advantages would be unprecedented. Genetic enhancement has the potential to produce far greater personal benefit than other privileges, such as a superior education, or great wealth. Now, even the most advantaged person cannot overcome genetic limitations. Even the best education can do only so much for someone who is not "gifted." If traits can be improved through genetic manipulation, however, "natural" or "innate" limitations no longer need be obstacles to personal success.

Genetic enhancement also would create demand for manipulating the genetic characteristics of germ cells to give enhanced traits a better chance of being passed on to succeeding generations. In Chapter Two, we discussed germ cell therapy for genetic disorders, and concluded that somatic cell therapy—or gene therapy targeted at affected tissues of the body—was more likely to be pursued than germ cell therapy—that is, manipulating the genetic structure of reproductive cells, namely, eggs and sperm—because somatic therapy would provide essentially the same benefits with fewer difficulties. But we also observed that germ cell manipulation would have greater appeal in the case of genetic enhancement. This stems, in part, from the likelihood that germ cell enhancements would be more effective than somatic enhancements and also from the desire to pass enhanced traits on to future generations. Genetic enhancement, then, may confer its advantages in perpetuity.

To summarize, the foregoing discussion demonstrates that genetic technologies will have enormous influence, both positive and negative, on people's lives and that the nature of the impact will depend on the access that individuals, or in some cases their parents, will have to these technologies. Most of the work that is being done on this subject focuses on the problems created when individuals encounter technologies that harm them. A great deal has been written, for example, about the risks of discrimination by employers and insurers against persons who have been tested for genetic disorders.[24] Relatively little attention, however, has been paid to the impact of beneficial technologies on those who are given access to them and on those to whom access is denied. It is to this subject that we now turn. And the first question that we must ask is: what will determine whether or not a person obtains access to beneficial genetic technologies?

5
Access to Genetic Technologies

As we saw in Chapter Four, the allure of genetic technologies will create high demand for them. People will want information about their
genetic endowments so that they can make important life-style and
reproductive decisions and so that they know whether to seek preventive or mitigative treatment for genetic disorders. They will seek gene
therapy as the definitive preventive measure or treatment for many of
these disorders. They will embrace genetic enhancements when they
become available as a means of gaining socioeconomic advantage in
their competitive environments.

Given the benefits that people will obtain from genetic technologies, will they be able to satisfy their demand? Certain conditions must
be satisfied in order for someone to gain access to genetic technologies.
At the outset, we will explore two of them: there must be an adequate
supply, and people must be able to pay for them. Later, we will discuss
a third prerequisite: that people are informed about the technologies
and know to ask for them.

SUPPLY SHORTAGES

Society has encountered shortages of supply of medical services in the
past. In some cases, the shortage has been due to technical problems
that have limited the availability of the resource. A classic example is
shortage of penicillin after it was first discovered. The new wonder
drug was in short supply because of the slow growth rate of the penicillin mold and impurities in the production batches. The shortage
produced some odd allocation policies; perhaps the most notable was
the military's allocation of access to penicillin during the early stages
of the Second World War.[1] Since the goal of military medicine was to
return soldiers to service as quickly and efficiently as possible, soldiers
with venereal disease, rather than combat casualties, received the
drug; the former could rejoin their units immediately, while the latter
usually required additional medical care.[2] This practice ended only
when the manufacturing difficulties were solved and the drug began
to be mass-produced.

A current example of a shortage of supply is the lack of sufficient organs for transplantation. More than forty-four thousand people are currently awaiting organ transplants in the United States. In 1994 more than fifty-six thousand people died while waiting for an organ, including more than thirty-eight thousand waiting for kidneys and more than six thousand waiting for hearts.[3] This shortage problem is unlikely to be solved unless there is a dramatic increase in the rate of organ donation, or unless techniques are perfected that would allow the use of nonhuman organs.[4]

In some cases, supply shortages have not been due to technical barriers so much as to cost. This is best illustrated by the outcry against the allocation of kidney dialysis machines in the 1960s. Dialysis is a technique whereby blood is removed from patients whose kidneys are no longer able to remove impurities from the bloodstream. The blood is cleansed in a dialysis machine and is then pumped back into the patient.

Early efforts to dialyze patients successfully were hampered by technical difficulties, such as the need repeatedly to puncture the patient's blood vessels, which the vessels could not long withstand.[5] In 1960, however, a doctor named Belding Scribner developed a tube or "shunt" that could be attached permanently to a patient's vein, thus avoiding the need to repuncture the patient, and that had a valve that could be hooked up to the dialyzer.

Scribner's shunt created an enormous demand for dialysis machines, which cost several thousand dollars apiece, and for the technicians to operate them.[6] Hospitals did not have enough machines to provide them to everyone in need. Wealthier patients paid their physicians to buy machines for them and hire technicians,[7] but poorer patients continued to die without dialysis treatments.[8] The shortage was only resolved when Congress, reacting to mounting public pressure,[9] created the End-Stage Renal Disease Program in 1972, and extended Medicare coverage of dialysis to persons in need even if they were under the age of sixty-five.

Given the early developmental stage of human genetic technologies, it is difficult to predict whether manufacturers will encounter any production difficulties that will lead to significant problems of supply. In the short term, however, shortages may occur for two reasons. First, when they are initially introduced for human use, genetic technologies will be regarded as experimental. Generally, the only people who gain access to experimental medical technologies are those who are enrolled in carefully controlled investigations to determine the safety and efficacy of the experimental product or technique.

Lack of access to experimental technologies might not seem to be detrimental to patients because, in theory, an experimental treatment is one that does not have proven patient benefits. Indeed, as we mentioned in Chapter Four, ethicists in the past have seemed more concerned that people might be coerced into becoming experimental subjects than that they would be denied access to experimental treatments. On the other hand, experimental treatments may be the only hope for patients with otherwise incurable diseases and conditions, particularly those that are life-threatening. This is illustrated by the demand for access to experimental drugs by persons with AIDS. If the only people who obtained access to experimental genetic technologies were those enrolled in clinical investigations, many people who believed that they might benefit from the technologies would be denied access. In this sense, then, experimental status causes a shortage of supply.

The U.S. Food and Drug Administration (FDA), which oversees the testing and approval of experimental drugs and medical devices, recently has taken steps to provide greater access to experimental treatments for life-threatening diseases such as AIDS. The agency has allowed manufacturers to distribute drugs broadly to patients before all required testing has been completed.[10] Between 1987 and 1993, twenty-eight drugs were made available under this program.[11] This policy may alleviate some of the shortage of experimental genetic technologies, but it is unlikely to affect technologies for non-life-threatening disorders or genetic enhancements.

A second limitation on the supply of genetic services may be a lack of health care professionals trained in their delivery. A shortage of genetic counselors is currently believed to be a major reason why more people do not receive genetic counseling.[12] One way to alleviate this type of shortage would be to permit services to be rendered by less well-trained persons. Genetic counseling might be performed by trained lay counselors. But the lack of training may cause a decline in the quality of the services, which might cause patient harm.

If the supply of genetic services is likely to be less than the demand for them, it may be useful to examine how we have allocated access to other medical treatments that have been in short supply. We already have mentioned penicillin. Another more current example is intensive care unit (ICU) beds in hospitals. There may not be enough of these beds for all of the patients who doctors determine require intensive-unit care. When this happens, a hospital has to decide which patients will be given priority. Sometimes this decision results in transferring intensive care patients to beds on general wards with less advanced life-support capabilities, in order to make way for other patients.

Patients appear to be selected to receive intensive care primarily on the basis of medical factors, such as the likelihood of their expected survival or the severity of their illness.[13] Medical factors also play a large role in the allocation of transplant organs. Unlike decisions about which patients will gain access to ICU beds, which are made according to the policies of individual hospitals, the rules for distributing transplant organs are prescribed by a national organization sanctioned by federal law—the United Network for Organ Sharing, or UNOS.[14] According to the UNOS rules, patients must be placed on waiting lists and ranked according to the seriousness of their illness, the likelihood that the transplant will succeed, and the length of time they have been waiting. The only other legal basis for preferring one patient over another is geographic location: patients who are in close proximity to donor hospitals are given preference for certain transplant organs over patients who are farther away. This stems from the fact that some organs cannot withstand the delay occasioned by transporting them or the patient a long distance.

The examples of ICU beds and transplant organs suggest that medical factors could be a key method that will determine which individuals will be given access to genetic technologies that are in short supply. But this does not mean that there is only one clear answer to the question of who will end up receiving them. Medical factors are not that clear-cut. Emphasizing different medical factors, for example, will produce different results. Hospitals used to give priority to the sickest patients in allocating scarce ICU beds. Then they switched to giving priority to the patients most likely to survive. (The percentage of patients dying in the ICU thereupon dropped from 80 to 20 percent.)[15]

In addition, physicians may disagree on which patients satisfy the medical criteria. In one study in the United Kingdom, for example, twenty-five kidney specialists were asked to use medical criteria to select thirty out of forty patients to receive dialysis. Only thirteen of the forty patients were chosen by all of the doctors, and every patient received at least one vote.[16] This suggests that patients who are denied a medical service by one physician or hospital on the basis of medical criteria would be wise to try to obtain a better result at another institution. The organ transplant program at the University of Pittsburgh, for instance, often accepts patients who have been rejected by other transplant programs.[17]

Even if access to scarce genetic resources was to be provided solely on the basis of medical factors, it may be impossible to prevent other factors from affecting the decision. For example, it has been reported that doctors rate the chances for the success of medical treatments higher in patients who resemble the doctors in terms of social and economic status than in other patients.[18] Informally, doctors are likely to put sympathetic patients at the top of waiting lists for organs, ICU beds, and other scarce medical resources. One way this can happen in the case of transplant organs is for doctors to describe favored patients as sicker than they really are in order to move them up on the waiting list.

Even though nonmedical factors may creep into allocation decisions surreptitiously, the American public is highly averse to allocating scarce medical resources on the basis of judgments concerning the patient's value to society. The prime illustration was the reaction to the dialysis programs in the 1960s, before Congress extended Medicare coverage to patients with end-stage kidney disease. In a 1962 article in *Life* magazine, Shana Alexander revealed how one hospital, the Seattle Artificial Kidney Center, selected the patients to receive access to the few dialysis machines that the hospital possessed. Among other things, the hospital relied on ". . . age and sex of patient; marital status and number of dependents; income; net worth; emotional stability, with particular regard to the patient's ability to accept treatment; educational background; the nature of occupation; past performance and future potential; [and] the names of people who could serve as references."[19] Not surprisingly, the center's decisionmakers became known as the "Seattle God Committee."[20] They were reported to give preferences to people who had been Boy Scout leaders and Sunday school teachers.[21] In the words of one pair of commentators, these practices ruled out "creative nonconformists, who rub the bourgeoisie the wrong way but who historically have contributed so much to the making of America. The Pacific Northwest is no place for a Henry David Thoreau with bad kidneys."[22] These and similar revelations outraged the public and were probably instrumental in persuading Congress to provide dialysis to patients under Medicare regardless of their age.

It is interesting to note that the establishment of the Medicare kidney disease program has produced a striking change in the makeup of the patient population that receives dialysis. Prior to the creation of the program, only 7 percent of those who received dialysis were African-Americans; afterwards, the percentage rose to 34.9 percent. The percentage of dialysis patients who had a junior high school education or less rose from 10 to 28.7 percent; the percentage of patients who were separated, divorced, or widowed increased from 5 to 25.2 percent; the percentage of women receiving dialysis increased from 25 to 50.8 percent.[23]

Another indication of the public's antipathy toward giving preferences to people based on their social status occurred in 1993, when Pennsylvania Governor Robert P. Casey received a heart and liver transplant less than twenty-four hours after being diagnosed with a need for the twin transplants. In contrast, the average patient at that time waited sixty-seven days for a new liver and 198 days for a new heart.[24] Reports in the press that Casey had jumped the queue because of his office led to an investigation by transplant officials.[25] The investigation concluded that hospital officials had not violated any allocation regulations since there were none that applied specifically to multiple-organ transplants, and that the governor had not "bumped" anybody who had been waiting longer for both organs.[26]

Although the dialysis experience and the Casey story reflect a deep-seated resistance toward giving patients preferences based on their social attributes, this approach still has adherents, at least in extreme cases. For example, a debate currently is being waged over whether murderers on death row should be given organ transplants. The Washington state legislature is considering a bill that would ban such transplants.[27]

A similar controversy arose over a proposal that patients who had contracted cirrhosis of the liver due to alcoholism should be placed lower down on waiting lists for life-saving liver transplants than patients whose need arose "through no fault of their own," even if the alcoholism patients had been abstinent for a long period as required by transplant center eligibility rules.[28] The federal government announced that Medicare would not discriminate against patients with alcoholic cirrhosis so long as they met the transplant centers' abstinence criteria and had a sufficient social support network.[29] In fact, the Department of Health and Human Services later held that giving a preference to nonalcoholic cirrhosis patients in need of liver transplants would violate the Americans with Disabilities Act.[30]

Another indication of what may be a greater willingness to allocate medical services on account of social factors is the passage of Proposition 187 in California, which prohibits the state from providing nonemergency medical care to illegal aliens.[31] The growing movement for welfare reform, viewed by many as being aimed at racial minorities, also may lead to cut-backs in eligibility for government-subsidized health care.[32]

Will society allocate scarce genetic technologies on the basis of social factors? Eugenics advocates who see genetic technologies as a way to "perfect" the human gene pool might push for allocation policies that favor persons with relatively poor genetic endowments. These policies also might appeal to liberals as a means of rectifying gross disparities of social status by giving persons of low status access to genetic enhancements that promote upward social mobility. On the other hand, some people may prefer to reserve scare genetic resources for those who are gifted with superior genes, on the theory that they would make the best use of them, thereby yielding the greatest amount of social good.

Even if society officially eschews an individual's social attributes as a consideration in allocating access to genetic technologies, a person's socioeconomic status may still play a key role. For there is a second condition that an individual must satisfy in order to gain access to genetic technologies: he or she must have insurance or be able to pay for it personally. The ability to pay may combine with actual shortages to limit access to a medical treatment. For example, people are not placed on a waiting list for an organ transplant unless they can show proof that they or their insurance will pay for it. An examination of our health care system shows that, regardless of how unpopular it may be to give people preferences on the basis of their social status, we are willing to tolerate a significant degree of unequal access to health care based on wealth.

INSURANCE COVERAGE

By far the most important factor governing access to health care in the United States is whether an individual has public or private health insurance. Lacking insurance, individuals would have to depend on their own private resources to purchase genetic services, or on charity to provide the services for free; neither source may be available.

Currently, health insurance is a patchwork of private and government plans. Private insurance includes both individual and group policies sold by commercial insurers and self-insured plans offered by employers to employees and their families. Altogether, private insurance paid for 31 percent of all personal health expenditures in 1995.[33] The principal government programs are Medicare and Medicaid. In 1995 they accounted for an additional 33 percent of all health payments. Only about 19 percent of health expenses in 1995 were paid by individuals out of their own funds.[34]

At least 41.9 million Americans lack health insurance at any one time.[35] Approximately ten million of these are children.[36] In 1995, an additional twenty million Americans lacked health insurance for only part of the year. In 1994, 35 percent of the uninsured under age 65 had incomes of less than fourteen thousand dollars a year; an additional 30 percent had annual incomes under twenty-five thousand dollars. While in some cases, these uninsured persons are able to pay for care themselves or to obtain charity care, in other cases they go without health care. Studies confirm that people without insurance get less health care than do people with insurance.[37]

Many people lose their health insurance when they lose or change their jobs. This is becoming a particularly acute problem because most people with private insurance obtain it through their employer, and employers lately have been downsizing their workforces. In many cases, people who change jobs may become insured by their new employer, but they lose coverage for health care problems that they had when they switched jobs—so-called "preexisting conditions."[38]

Until recently, there was a fair amount of charity, or what is sometimes referred to as "uncompensated" health care. Actually, health care providers such as hospitals and physicians usually do receive compensation for these services, but they receive it indirectly, by adding the charges for these services to the bills submitted to paying patients or their insurers.[39] This practice is known as "cost-shifting." Cost-shifting has not been limited to covering the costs of health care provided to the poor. It also has allowed health care providers to offer discounts to patients enrolled in large groups, such as HMOs, and to make up the difference in revenues by charging more to other patients. In 1994, for example, a hospital in Washington, D.C., charged a patient who was insured under a traditional union health insurance plan $28,113 for a triple coronary by-pass operation, while it charged a patient enrolled in Kaiser Permanente, a large HMO, only $10,987 for the exact same procedure.[40]

The problem for health care providers is that cost-shifting is becoming more and more difficult as health care providers slash prices in order to compete and as insurers and employers become less willing to subsidize medical care for people who are neither their insureds nor their employees. More and more patients are joining or being forced into HMOs, which insist on obtaining discounted rates from providers. Fewer and fewer patients have insurance plans willing to pay full fare, including the bill for the uninsured. The patient described above, whose union benefits plan paid an additional seventeen thousand dollars for his by-pass operation in 1994, is now enrolled in an HMO.

Government health insurance programs are not expanding to take up the slack. In fact, the outlook is for cutbacks in Medicare and Medicaid. For example, Congress is considering increasing the amounts that Medicare beneficiaries themselves must pay for covered services. Federal subsidies for state Medicaid programs, currently comprising between 48 and 75 percent of state Medicaid budgets, currently are tied to certain minimum standards for eligibility. Yet Congress may transform these subsidies into block grants that would free the states to adopt their own eligibility criteria. Given the overall pressures on state budgets, this new freedom almost certainly would result in fewer people being eligible for Medicaid assistance.

The plight of the uninsured, coupled with the rising costs of health care and the increasing unwillingness of insurers to pay the costs of care for others, prompted President Clinton to propose legislation that would have guaranteed health insurance for virtually everyone. This effort collapsed under pressure from health insurers and others, and there are no plans to revive it. Given current trends, then, many people are destined to be denied access to genetic technologies because they do not have health insurance.

Even if a person is insured, however, the insurance program may not pay for, or "cover," every service that the person seeks. Many private insurance plans, as well as public programs such as Medicare, do not cover certain categories of services such as prescription drugs, infertility services, or treatments considered to be experimental. If a service is not covered, even a patient who is insured will not be able to obtain it unless the patient can afford to purchase it with private funds or can get it for free.

The importance of coverage policy was clearly illustrated in the debate over national health reform. President Clinton's proposal, the Health Security Act, would have extended health insurance to almost everyone, but it would not have covered all potentially beneficial medical services. People would only have been guaranteed access to a package of basic health benefits.

The basic benefits package was quite generous, but the description of covered services in most cases was very general and left a great deal of room for government discretion. The President's bill stated that the basic benefits package would include inpatient and outpatient hospital care, services of physicians and other health care professionals, preventive services, mental health and substance abuse services, home health and some long-term care, family planning and pregnancy services, hospice care, laboratory, radiology, and diagnostic services, prescription drugs, rehabilitation services, durable medical equipment and prostheses, some vision and dental care, and some of the costs associated with participating in clinical trials.[41] The plan described in considerable detail the mental health and substance abuse services and the specific preventive services that would be available to persons in specific age groups. However, it provided no explanation for what was meant by "family planning services and services for pregnant women"—a category of services with special significance for genetic technologies—beyond stating that they included prescription contraceptive devices and that family planning services would be covered if they were "voluntary." The legislation was silent on whether or not abortion would be included, an issue that was the subject of substantial controversy between the administration and its critics, and one never fully resolved before the administration abandoned the effort to achieve national health reform.

In the absence of national health reform, Medicare and Medicaid will continue to be the major public health insurance programs. These programs have their own coverage limitations. The Medicare statute precludes payment for routine physical examinations, cosmetic surgery, routine eye and dental care, self-administered prescription drugs, hearing aids, and long-term nursing home care.[42] In addition, the law prevents Medicare from paying for "items or services which . . . are not reasonable and necessary for the diagnosis or treatment of illnesses or injury or to improve the functioning of a malformed body part." The federal agency that administers the Medicare program—the Health Care Financing Administration—is responsible for enforcing these exclusions. To a large extent, it delegates coverage decisions to private companies—typically, insurers like Blue Cross/Blue Shield—that contract with the government to review Medicare claims. But the Health Care Financing Administration occasionally is called upon to issue program-wide coverage directives, stating what it will and will not pay for. In preparing these directives, the agency consults medical experts and other branches of the government, such as the Food and Drug Administration. The agency publishes the directives in the *Federal Register* and collects them in a manual that it distributes to Medicare providers.

Coverage policies issued by the Medicare program used to have greater significance before the adoption in 1984 of a new method for paying hospitals for treating Medicare patients. Under the old approach, a hospital billed Medicare for each service it provided and was paid more, the more services it gave to patients. The new system, called the "diagnosis-related group," or DRG approach, pays a hospital essentially the same amount, depending on the patient's diagnosis, regardless of what services the hospital actually provides. This gives hospitals an incentive to provide as few services as possible, since the less they provide, the less it costs them and, therefore, the greater the difference between what they spend and the DRG amount that they are paid, i.e., the more profit they make. Since hospitals already have an incentive to restrict access to services under this system, particularly access to expensive services that provide little or no known patient benefit, there is no longer much reason for Medicare to dictate which specific technologies it refuses to pay hospitals for providing; if anything, rather than being concerned that hospitals will provide too many services to patients, the government needs to monitor the quality of patient care to make sure that hospitals do not provide patients with *too few* services. The main role for Medicare coverage restrictions has become one of oversight, making sure that hospitals are not admitting patients and getting reimbursed by Medicare when the *only* reason for the admission is to provide the patient with a treatment that Medicare does not cover—for example, an experimental organ transplant.

Medicaid, the federal/state program that provides health care for certain categories of the poor, also operates under certain coverage restrictions. Some of these limitations are contained in federal legislation. For example, the federal government will not help state Medicaid programs pay for cosmetic services, routine dental and eye care, or abortions, except when an abortion is necessary to save the life of the mother or when the pregnancy results from rape or incest.[43] Other coverage restrictions are imposed by the states themselves. For example, many states do not pay for liver transplants for cancer patients, patients with active hepatitis B, or patients who are more than seventy years old.[44]

Private health insurance coverage is governed largely by the terms of the insurance policy itself. Private health insurance benefits may be subject to state laws or "mandates" that require insurers to provide certain benefits as part of their benefits package. For example, a number of states require insurers to pay for mental health and substance abuse treatment services. In the wake of reports that insurers were limiting maternity stays in hospitals to as little as eight hours, several states have enacted minimum maternity stay requirements.[45]

Instead of purchasing health insurance from a commercial company for their employees, employers may choose to provide their own insurance. These so-called "self-insured" plans typically are exempt from state mandates and other state insurance laws under the federal Employee Retirement Income Security Act, or ERISA.[46] Under ERISA health plans, employers generally are free to offer whatever health benefits they wish, subject only to two limitations. First, under federal law, they may not deny benefits in such a way that they discriminate against the disabled.[47] As we will see later on, this prohibition may significantly affect both private and public limits on genetic technologies. Second, employees may pressure employers to agree to offer certain health benefits to their employees, particularly if the employees are organized into unions, since health benefits usually are an important part of collective bargaining agreements.

Private health insurance policies, including those issued by employer self-insured plans, typically describe what they cover and what they exclude from coverage, using a combination of general and specific provisions. A general provision might say that the policy covers "all medically necessary physician services," and that the policy excludes "experimental" treatments. Specific provisions describe particular covered and excluded services, such as infertility treatments or cosmetic surgery.

Controversies may arise between the insurer and insureds over what services the policy will pay for. This is particularly likely if the policy does not mention a medical service specifically. The most common controversy arises over what services the policy excludes as "experimental," since often the policy will not list specific excluded experimental services. Currently, for example, many insurers and insureds are arguing over whether a certain treatment for cancer— high dose chemotherapy coupled with autologous bone marrow transplantation (HDC/ABMT)—is experimental.[48] Typically, the insurance policy will not specifically list this treatment as an excluded service, but will state merely that the policy excludes experimental treatments. If HDC/ABMT is deemed experimental, then the insurer does not have to pay for it. A number of these disputes have ended up in court, with patients arguing that many doctors regard the treatment as beneficial and, therefore, that it should not be regarded as experimental. Insurers argue for their part that the information upon which patients are relying is not scientific enough and that more studies are needed before insurers should have to pay for the treatment. In many cases, judges have sided with the patients. Insurers react by rewriting their policies to include a more specific list of excluded experimental treatments.

This raises the question of why insurers do not simply include an extensive, specific list of excluded treatments in their policies in the first place? The explanation stems from a principle that courts use to interpret agreements such as insurance policies. If the agreement contains a general term, such as an exclusion for treatments that are "experimental," plus a list of specific items that fit the term, such as HDC/ABMT, then the courts are likely to decide that specific items that are not on the list—such as new treatments that have come along since the last time the insurer updated the policy—are not excluded by the general term. This forces insurers to change their policies almost constantly, which is difficult or impossible under their arrangements with insureds, or to pay for services that they regard as experimental but that are not on the exclusion list. The alternative is to continue to use general exclusionary language and to hope that the courts begin to side more with insurers.

We have seen so far that whether or not an individual will obtain access to medical care depends not only on whether the individual has health insurance, but on what the insurance covers. In order to understand how these factors may influence access to future genetic technologies, we must introduce another key theme that animates health care policy: the need to control health care costs.

With health care spending rising at more than double the rate of inflation, a major impetus behind national health care reform has been the need to contain health care costs. One of the primary techniques that private and governmental third-party payers have used to control health care costs has been to reduce access to insurance. Employers eliminate insurance as an employment benefit, or severely limit the amounts that they will pay. Commercial health insurers refuse to insure high-risk patient populations, and even when they do provide insurance, they exclude "preexisting conditions," which denies meaningful access to insurance to the chronically ill and to those who change jobs. Limiting access to insurance also has been used to hold down the costs of the Medicaid program. In order to receive federal Medicaid funds, states must provide third-party payments on behalf of certain categories of patients—chiefly, parents and children who qualify for Aid to Families with Dependent Children, and certain persons who are aged, blind, or disabled. In addition to meeting this so-called "categorical" requirement, Medicaid beneficiaries may not have incomes higher than the federal poverty level, which was $12,324 for a family of three in 1994.[49] If states provide Medicaid to persons with higher incomes, they will lose their federal subsidies. But there is nothing to prevent states from *lowering* the Medicaid income threshold below the federal poverty level. In the past, if states could not come up with sufficient funds to finance their Medicaid programs, they simply lowered the income threshold for eligibility, thereby reducing the number of eligible individuals. Under the Texas program, for example, categorically eligible Medicaid beneficiaries cannot earn more than 18.6 percent of the federal poverty level for a family of three, a whopping $2,292 per year.[50]

This classic technique of limiting the persons who have access to health insurance has clashed with the health reformers' goal of extending health insurance to everyone. This leaves only a few alternatives to control health care costs. One is to reduce the proportion of health care expenses that insurance pays for—for instance, by increasing the size of patient copayments or deductibles. But this may preclude persons with low incomes, who cannot afford the coinsurance payments, from obtaining needed health care services. Another alternative is to reduce the amounts that health care providers are paid. But this reduces access by driving providers away from patients who cannot supplement meager government payments with private insurance or personal funds. For example, many doctors no longer will treat Medicaid patients because of the level of reimbursement; the average amount that they are paid is only 56 percent of what Medicare pays for comparable services. Only one alternative remains: to limit the services for which health insurance pays.

The interplay of these factors was illustrated most clearly in Oregon's efforts to modify its Medicaid program. In 1989 only about 160,000 of the 300,000 Oregonians with incomes below the federal poverty level were eligible for Medicaid. The rest either did not fit into one of the categories that Congress had established for Medicaid eligibility—for example, they were childless individuals—or their incomes exceeded the state's eligibility threshold, which at the time was 58 percent of the federal poverty level. Under the leadership of a physician who was president of the state senate, the legislature responded by broadening Medicaid eligibility to include all state residents whose incomes were below the federal poverty level.[51]

The problem was how to pay for this sudden doubling of the Medicaid population. Obviously, the state could not avail itself of the traditional method for dealing with an increase in Medicaid costs, which was to reduce the number of people eligible to receive benefits. One option was to raise state taxes. But voters were not about to accept a tax increase; in fact, the Oregon legislature, following in the footsteps of California Proposition 13, had just enacted a rollback of state property taxes, which were used to finance public education. The lost property tax revenues for education were to be replaced by funds from the state's general tax revenues.[52] This meant that any additional funding for an expanded Medicaid program, which also would come out of general revenues, had to compete with funding for schools. The state also could not hope to finance an expanded Medicaid program—or anything else—by creating a budget deficit, a technique used by the federal government; the Oregon state constitution required the budget to run in the black.

As mentioned earlier, one way to broaden eligibility without increasing costs might be to reduce the amounts that health care providers were paid for treating Medicaid patients. But Oregon already had reduced Medicaid reimbursement to 55 percent of what providers were paid for furnishing comparable services to Medicare patients.[53] Policymakers feared that limiting reimbursement further would discourage physicians and other providers from agreeing to accept the new Medicaid enrollees as patients and would threaten the quality of care that Medicaid patients would receive.

Oregon took the only remaining approach. It broadened eligibility by reducing the medical services that Medicaid patients were entitled to receive. It did this by embarking on an ambitious, unprecedented effort to devise a method for publicly rationing health care. It appointed a Health Services Commission, which included five physicians, a public health nurse, and a social service worker, as well as four consumer representatives.[54] This group set about creating a list of 709 health services, expressed in terms of "condition-treatment pairs" like "antibiotics for pneumonia" or "surgery for appendicitis." These were ranked in order of "importance." The rankings resulted from a complex process, overseen by the commission, that included surveying state residents to determine which health states they valued the highest, estimating the likelihood that various medical services would produce these health states in patients with various conditions, and finally, permitting the commissioners to adjust the results based on their subjective views.[55] When the list was completed, actuaries took it and, starting from the highest ranked services at the top of the list, estimated how much it would cost to provide each service, plus all those ranked above it, to the expanded Medicaid patient population. This compilation was given to the legislature, which determined how much it wanted to budget for the Medicaid program as a whole. A line was then drawn on the list of services at the point at which the cumulative cost of the program would equal the amount the legislature had decided to spend. All services above the line would be reimbursed; services below the line would not. This meant, essentially, that they would not be available to Medicaid patients.

From the outset, the Oregon plan encountered difficulties in calculating the rankings. What sounded good in theory initially produced results that seemed counter-intuitive. Life-saving heart, liver, and bone marrow transplants, for example, ended up lower on the list than smoking cessation programs, foot care for the elderly, and dentures.[56] The commission responded by revising the ranking process. The new list of 709 condition-treatment pairs was still criticized for being based on inaccurate information about costs and benefits, and for producing clinically questionable results.[57] Perhaps the most serious objection was to the changes that the members of the Health Services Commission, acting on their subjective judgments of reasonableness, made in the rankings. This caused 25 percent of the condition-treatment pairs to move up or down from their initial position on the list, with the result that almost all of the 709 pairs changed their original ranking. Many moved as much as 100 places.[58] Many felt that the commissioners' exercise of discretion deprived the ranking process of much of its objectivity. The commissioners defended their actions as necessary in order to reflect shared societal values.

The proposed ranking of liver transplants for cirrhosis provides an example of the effect of the commissioners' exercise of discretion on the ranking process. The initial ranking process placed liver transplants at number 263. Then, the commissioners split number 263 into transplants for alcoholic cirrhosis and transplants for nonalcoholic cirrhosis. Based on the perception that persons in need of a liver transplant because of their excessive consumption of alcohol were less deserving, the commissioners moved transplants for alcohol-related cirrhosis down to number 690. When the Oregon legislature drew the coverage line, it fell at 587, which meant that only nonalcoholic transplants would be covered under Medicaid.

The Oregon plan also was criticized as unfair to the poorest of the poor. The thrust of the proposal was to ration medical services that previously would have been provided to Medicaid patients so that more people would be eligible for health care benefits under Medicaid. This sounds laudable until one realizes that the people who were to lose services were those Oregonians who already were eligible for Medicaid, meaning, among other things, that they had incomes of less than 58 percent of the federal poverty level, while those who were going to gain eligibility comprised a substantial number of persons with higher incomes.

What finally proved fatal to Oregon's original plan was largely unanticipated—the realization that it would violate federal laws prohibiting discrimination against persons with disabilities. Following objections by pro-life groups and advocates for the disabled,[59] Secretary of Health and Human Services Sullivan ended up blocking the Oregon effort for three reasons. First, the priority list ranked services based in part on the quality of life of the patient following treatment. This meant that treatments for persons with disabilities that left the person still disabled would be ranked lower than the same treatment for persons who would be left without disabilities. Second, Oregon measured quality of life according to the attitudes of the general population rather than those of persons with disabilities, which could cause disabled conditions to be undervalued. Third, the administration objected that the low ranking given certain specific treatments, such as liver transplants for alcoholic cirrhosis and life support for extremely low-birth-weight infants (less than five hundred grams) of less than twenty-three weeks gestation, were discriminatory on their face.[60] In a sweeping statement, Secretary Sullivan, citing an earlier Supreme Court case, *Alexander v. Choate*, concluded that if Oregon wanted to proceed with its rationing scheme, it must reformulate its priority list to employ only "content neutral factors" that did not take disability into account and that did not have a particular exclusionary effect on persons with disabilities.

Oregon went ahead and recalculated its list, using a highly simplified approach. It now ranked treatments on the basis of the relative probability that a patient would be "symptomatic," "nonsymptomatic," or deceased in five years and eliminated comparisons based on the patient's quality of life, except in cases of ties between treatment rankings.[61] The revised list was sent to the newly elected Clinton administration for approval. It was once again rejected on account of the disability discrimination laws. This time, the problem was Oregon's use of quality-of-life comparisons to break ties. Oregon once again recalculated its list, avoiding any consideration of quality-of-life. Treatments were compared initially based only on the relative probability that the patient would be alive in five years; ties were broken based on relative cost and, if necessary, on the alphabetical order of the treatments.[62] This list eventually was put into effect.[63]

Oregon's difficulties made it unlikely that other states would adopt its approach. But in the meantime, another development has taken place that is a more probable blueprint for the rules that will govern access to health care services in general and genetic services in particular: the widespread use of "managed care" to contain costs.

Managed care can take a number of forms. Third-party payers, such as employers or health insurance plans, can establish financial incentives to reward physicians and other health care providers for withholding care. The providers may continue to be paid in the traditional fee-for-service manner, receiving a separate fee for each service or unit of service (such as a day of hospitalization) they provide. In contrast to the traditional payment scheme, however, managed care plans reward the provider financially for holding down the amount of services furnished patients and hence the third-party payer's costs. In a common type of incentive program, the managed care organization withholds a portion of the provider's fee and refunds it to the provider at the end of the year if the provider remains within the managed care organization's utilization guidelines. This form of managed care often includes a "gate-keeper" feature in which the financial incentives are focused on a patient's primary care physician. This physician is given the responsibility of deciding what additional services the patient will receive; for example, whether the patient will be referred to a specialist or admitted to a hospital.

Managed care also includes a broad range of other efforts to influence resource allocation decisions made by health care professionals. Many of these approaches are embraced within the term "utilization review." This encompasses requirements that the professional obtain permission from the third-party payer, either before a reimbursable service is rendered (prospective utilization review) or before the payer will pay for a service that already has been provided (retrospective utilization review). For example, prospective utilization review often is required before a primary care physician may refer a patient to a specialist. Utilization review also includes more subtle forms of pressure to limit services, such as giving the provider data comparing the provider's service patterns with those of other providers or with preestablished guidelines, or notifying a provider that a utilization guideline is about to be or has been exceeded. The ultimate threat from a managed care organization is that a health care provider who does not limit services will be excluded from the organization's network of providers. Since the provider may depend on the network for a substantial source of patients, exclusion may be a serious and even fatal blow to the provider's practice.

All of these managed care approaches share one thing in common: they all aim to reduce services to patients in order to reduce costs. Ideally, the only services that are withheld are wasteful services—those providing little or no patient benefit. But after all the waste is gone, managed care continues to exert relentless pressure on health care providers to trim costs. At this point, the only services that remain for the provider to cut are those that provide important patient benefits.

Advocates of managed care might disagree that its primary aim is to reduce services. The objective instead, they might assert, is to provide relatively inexpensive services, such as preventive care, that reduce the need for expensive treatments later on. No rational managed care provider, they might insist, would deny care knowing that this would increase costs later on.

These claims do find some support in studies on managed care organizations. There is no question that HMO enrollees receive more preventive care than enrollees in traditional health insurance plans.[64] Moreover, if managed care providers skimped on necessary services, we would expect to see a deterioration of quality, yet studies generally show that HMO patients receive care of the same or better quality than other patients.[65]

These results must be viewed with caution, however. For one thing, it is too soon to expect to see clear evidence that managed care entities curtail access to beneficial medical services. The medical system is still so filled with waste, in the form of unproven or nonbeneficial interventions, and still maintains a sufficiently large margin of safety, that managed care providers can reduce costs substantially without significantly compromising the overall quality of patient care. But this cannot go on forever. Eventually, competitive pressures are bound to force managed care providers to save money by withholding medical services with established patient benefits. These same pressures will make it difficult for managed care providers to continue to expend money up-front, such as by providing more costly forms of preventive care or early treatments, in order to save money later on. The need for start-up companies to establish a foothold in the market and for older entities to expand their market share will encourage them to cut short-term costs at the expense of long-term savings. These pressures will be magnified in the case of for-profit providers, whose investors tend to demand short-term profitability.

Finally, the incentive to provide preventive services disappears once prevention has failed. Even if managed care providers were willing to make short-term expenditures to reduce their long-term costs, they would have little to gain financially from providing services to enrollees who became chronically or seriously ill and required extensive, costly care.

A related change that is taking place in the way health care is delivered and that will have an important impact on access to genetic services is the growing popularity of a method of paying for health care called "capitation." Under this approach, the payer (typically, an employer or insurance company) pays the health care provider a fixed amount each month for each person who may receive services from the provider. The provider receives this amount whether or not it provides services to the person. If it provides no services that month, the entire capitated payment sum, minus the provider's overhead, becomes profit. On the other hand, the more services the provider gives to the patients, the more it costs the provider and, therefore, the less the profit that remains at the end of the month. A patient who requires a lot of expensive care may not only exhaust the provider's capitated payment completely, but may force the provider to dip into its own financial reserves to pay for the patient's care. In the ultimate form of capitation, a single health care provider—often the patient's primary care physician—is paid a fixed amount per month to take care of all of the patient's health care needs, including the care the patient receives from specialists and in the hospital.

Earlier we described how even a patient who has health insurance may not be given access to medical services if those services are not covered by the insurance program. Under managed care and capitation, there is no longer a need for the insurer to hold down costs by telling the health provider what services the insurer will and will not pay for. Instead, the capitated payment system itself creates the necessary incentives for the provider to hold down costs. In effect, coverage decisions now are made by the provider itself rather than by the insurer.

The growth of managed care and capitation does not mean that a plan's coverage policy—what the health plan will and will not pay for—is no longer important in managed care and capitated plans. Coverage policy simply takes on a new function. Instead of identifying the maximum services the patient can receive, the plan's coverage policy establishes the minimum. It becomes the means by which patients assert their right to obtain access to the services that they are entitled to receive under their plan.

This "affirmative" use of a plan's coverage policy by patients is illustrated by recent court cases involving a portion of the Medicaid program known as the Early and Periodic Screening, Diagnosis, and Treatment (EPSDT) program. This provides coverage for a set of medical services that Congress stipulated must be extended to all Medicaid recipients under the age of twenty-one.[66] In 1989 Congress amended the law governing EPSDT services[67] to require states to provide all medically necessary EPSDT services, even if those services are not otherwise covered under the state's Medicaid program.[68] The significance of this amendment pertains particularly to life-sustaining organ transplants. As we mentioned earlier, states are not required to cover most organ transplants under their Medicaid programs.[69] When several states attempted to deny coverage of organ transplants to Medicaid recipients under the age of twenty-one, however, patients' families sued, arguing that the denial of coverage would violate the foregoing EPSDT amendments. The courts agreed.[70] More recently, a court held that a state Medicaid program could not refuse to cover organ transplants for adults if it covered them for children, since that would violate another provision of federal law, which requires state policies for covering organ transplants under Medicaid to "treat similarly situated individuals alike."[71] This decision would seem to require states to cover organ transplants for all Medicaid recipients, since they must cover them for children under the EPSDT program.

Thus far, we have seen that whether or not a person will receive access to medical services in general depends on whether the person has health insurance and on what services the insurance covers. Furthermore, we have noted that pressures to reduce health care spending are resulting in cutbacks in the services people are entitled to receive, both under government and private health insurance programs. We will now see how these factors influence access to the specific types of genetic services described in Chapter Three.

Coverage of Genetic Testing

One set of genetic technologies described in Chapter Three was valuable for the information that it could provide. These include a number of different types of genetic tests that would be performed at different stages of an individual's life. Whether or not these tests will be widely accessible will depend on several factors. One consideration is the cost. It is virtually impossible at the present time to predict how much it will cost to screen individuals for all the genetic disorders for which they are at relatively high risk, but the cost is likely to be extremely steep, given the number of disorders and the large number of persons who might be at risk. When genetic tests become widely available to predict an inherited susceptibility to such common diseases as cancer and heart disease, for example, the people who might be tested for these diseases would comprise virtually the entire population. Even if the per-unit cost of testing were relatively small, the total cost of screening large numbers of individuals could be prohibitive.

Added to the financial costs of large-scale screening programs would be the social costs. These include the kinds of discrimination described in Chapter Four. They also include the costs to individuals of being given incorrect results. No test is 100 percent accurate; it may incorrectly identify an individual as having a genetic disease or predisposition when the individual does not (a result called a "false positive"); or the test result may be negative when the individual does in fact possess the disease or trait (a "false negative" result). A false negative result can lead the person who was tested to refrain from beneficial followup and treatment. A false positive result, on the other hand, could cause the individual to obtain unnecessary treatments, to make incorrect life-style decisions, and to suffer needless emotional trauma.

To avoid the disagreeable consequences of a false positive test result, a person who receives a positive result on an initial screening test may have to undergo further testing to confirm or disprove the accuracy of the initial result. A positive result on the initial test for HIV antibodies, the ELISA test, is routinely followed by a more definitive test, the Western Blot. Follow-up testing adds substantial costs. The ELISA test only costs approximately twenty-five dollars; the Western Blot, however, costs several hundred.[72] Depending on the prevalence of the condition being tested for, these costs can be enormous. For example, a common test for colon cancer, the stool guaiac, cost only about four dollars in 1975. But because of the relative rarity of the condition and the high rate of false test results, it would cost more than forty-seven million dollars to detect the final three cases of colon cancer in the population.[73]

These financial, social, and personal costs might lead us to restrict access to genetic screening tests. Public or private insurance could refuse to pay for them. Or, public and private insurers might decide to pay only for those tests that would detect ailments that could then be prevented or effectively treated. As mentioned in Chapter Four, for example, all states require some genetic screening tests to be administered to newborns. These are tests for ailments that can be treated medically or that can be prevented or mitigated by life-style changes, such as diet.

A policy of paying only for screening tests for treatable or preventable diseases finds support in a report by the Institute of Medicine's Committee on Assessing Genetic Risks. The committee recommends that states mandate screening only if, among other things, they provide treatment and followup for the affected individuals.[74] Such an approach also is consistent with the emphasis on prevention in managed care plans.

A difficult question arises, however, with a screening policy of this type. What constitutes treatment or prevention? Would the possibility of a modest improvement in health status following detection be sufficient to trigger screening for the disorder? What if the potential improvement would be significant, but there was only a slim chance that it would occur? What about abortion? Does it count as prevention? If so, then we would have to screen for any genetic disorder that could be detected in the fetus during the early stages of pregnancy. Depending on how we defined treatment and prevention, we could end up screening almost everyone for a very large number of genetic disorders.

The costs associated with such a broad screening program undoubtedly would require additional access restrictions to be imposed. Most likely, tests will be provided only to people who present a high risk of having the genetic disorder. For example, one particular genetic defect has been associated with a high incidence of breast cancer in Ashkenazic Jewish women.[75] Screening for the defect might be appropriate for this group, but not for others who do not have as high a rate of the disease. The Clinton reform plan took such an approach; genetic testing would only have been covered for high-risk populations.[76]

Limiting genetic testing to high-risk populations also is supported by ethical considerations.[77] Genetic testing in low-risk populations may cause greater harm than benefit. Any genetic test raises the possibility that its results may be misunderstood or overemphasized. The classic example was the screening programs for sickle cell trait in the early 1970s, discussed in Chapter Four. Some individuals who tested positive for the trait, it will be recalled, mistakenly assumed that they would contract the disease or that they invariably would pass it on to their children, when in fact they were merely carriers of the trait who did not have the disease and would only pass it on to their offspring if they conceived a child with another person who had the trait.

Another risk of broad-scale testing is the possibility of error. As noted earlier, genetic test results can err by purporting to find a genetic disorder when one is not present (false positive) or by missing a genetic disorder that is in fact present in the tested individual (false negative). The chances of a false result increase substantially when tests are performed in low-risk populations. These risks may be worth taking when the individual stands a good chance of having the genetic trait at issue. But someone at low risk might have so much less at stake that these dangers could not be justified.

Just as we might find it difficult to decide how much treatment or prevention justifies screening, we also might have trouble defining when a risk is high enough. Is a 1 percent chance of carrying a mutation that leads to breast cancer sufficient to justify screening? How about a risk of 1 in 1,000, or 1 in 10,000? Even if the probability is high that the individual will suffer from a genetic disorder, how severe must the disorder be in order for screening to be deemed worthwhile? Would we screen for a minor ailment such as a slight loss of hearing or vision? Or would we only screen for life-threatening diseases? How should the fact that a genetic disorder will not manifest symptoms until later in a person's life affect the determination of the magnitude of the risk? Current disputes over which populations should be given access to such nongenetic technologies as mammograms and pap smears, and how frequently these tests should be administered, illustrate some of the types of controversies that would arise. Not only are medical experts likely to disagree about how great a risk there must be to warrant screening, but there is likely to be disagreement within the target population.

One solution to the problem of deciding when to test might be to permit individuals to decide for themselves. That way, a person who does not like to take chances might opt to be tested for a lower risk of disease than would someone less risk-averse. Such an approach would be consistent with the principle of informed consent to health care, which holds that patients should decide which risks and benefits they choose to accept. Indeed, some people might decide to be tested for disorders that could not successfully be prevented or treated. Prospective parents might feel that it is important for them to know about and be able to prepare for a child with a genetic disorder even though the disorder could not be treated and the parents had decided not to obtain an abortion. Adults might want to know whether or not they harbor a genetic defect that would predispose them to an untreatable disease, so that they can make career, family, and life-style decisions. But the problem with this approach is obvious, given our original premise that we need to limit access to genetic tests to keep costs down. If people are allowed to decide for themselves what tests to obtain, they may opt to have too many. Costs once again could spiral out of control. In short, those people who will need to depend on health insurance in order to obtain access to genetic technologies will have to accept the coverage limits that the insurers impose.

In addition to the question of whether third-party payers will pay for genetic testing itself is the question of whether they will pay for the counseling that must accompany it. Counseling is critical. A person contemplating genetic testing needs assistance to identify those disorders for which he or she is at high risk. The individual will also need advice on which tests are available, how accurate they are, what the costs and benefits of the tests are, what the results mean, and what follow-up monitoring or treatment he or she requires. Counseling will become increasingly important in order for people to understand their options as the number of genetic tests grows.

Historically, third-party payers often have not paid for genetic counseling.[78] Even when insurers cover counseling, they pay relatively low rates for the necessary time-consuming office visits, as compared with their reimbursement rates for surgical and other high-tech procedures. If this practice should persist, it alone could severely limit access to genetic testing. As the number of genetic tests increases, the time to counsel patients also would increase, probably exponentially. Inadequate or nonexistent insurance coverage could lead people to obtain an inadequate amount of counseling, or to forego it completely. This could cause them to make ill-considered decisions about whether or not to be tested, and to misinterpret test findings.

This brings us to the third criterion for determining whether an individual will obtain access to genetic information services: even if there were an adequate supply and people had insurance or otherwise could pay for it, they will not obtain access unless they know to ask for it. One way of limiting access to genetic testing, therefore, would be to limit information about what testing is available.

This approach actually has been used in the past to limit access to medical services. The most notorious example was in Britain, and once again it related to kidney dialysis. Although Britain has a national health system, which in effect provides health insurance for everyone, at one time it did not pay for dialysis for persons over fifty-five years of age. To avoid being confronted with people demanding the life-saving treatments, doctors simply did not inform those kidney patients who were over the age limit that there was such a thing as dialysis, or that it could keep them alive.[79]

Surprising as it may seem, a similar approach currently is being attempted in the United States. Managed care plans are inserting "gag" clauses in their agreements with physicians. These provisions forbid physicians to tell their patients that they might benefit by obtaining health care services that the plan does not cover. Managed care plans may be motivated by several concerns. Like their counterparts in the British national health system, they may fear that if patients demand the services, the plan will be forced to cover them. In numerous court cases, judges have forced managed care plans and other health insurers to pay for treatments that the plans attempted to exclude from coverage. Most of these cases involve new treatments for cancer that have not been fully tested. Managed care plans also may be concerned that the patient may seek the services "out-of-plan." In some instances, the plan would still have to pay a portion of the out-of-plan costs. Finally, managed care plans want to keep their enrollees happy. Not telling someone about beneficial services that the plan does not cover may be seen as a way to prevent the enrollees from becoming disgruntled and switching to another health plan.

Failing to provide information to patients is clearly inconsistent with the principle of informed consent mentioned earlier. The practice of British doctors withholding information about dialysis occurred because British law did not recognize the informed consent requirement. In the United States, the courts are likely to strike down gag provisions in managed care contracts with physicians. It is interesting to note that, although the Oregon Medicaid program does not pay for all medically necessary treatments, it still requires doctors to inform patients about beneficial services that the program does not cover. [80]

Even if physicians in the United States would not voluntarily withhold information from their patients about potentially beneficial but expensive genetic testing merely because managed care plans told them to do so, there is a more diabolical technique that plans are beginning to employ that threatens to produce the same result. They are beginning to "capitate" physicians. This means that the plan gives the physician a fixed amount of money for each patient that the physician may treat. If the physician does not see the patient, or if the cost of treatment, including overhead, is less than the fixed or capitated amount that the physician receives from the plan, the physician makes a profit. If the patient is very ill and requires expensive hospitalization or high-tech care, the physician may be required to expend more resources and will lose money if he or she spends more on the patient than the capitated amount. This payment scheme could cause physicians to refuse to inform patients about expensive genetic tests or treatments that might benefit the patient, despite the fact that they would be violating the law as well as their code of professional ethics.

Coverage of Gene Therapy

Many of the same factors that would combine to limit access to genetic information services would also restrict access to gene therapy. As described earlier, gene therapies will be considered experimental when they are first developed. Generally, only those patients enrolled in scientific studies of the new techniques will gain access to them. Moreover, gene therapies may remain in experimental status longer than will genetic tests, in view of the more complex assessments of safety and efficacy that may be required. In contrast to genetic testing, which entails little more than withdrawing a sample of the patient's blood, gene therapies may cause adverse physical effects in patients.

These need to be detected and measured in clinical trials and then compared with the potential benefits from the therapy to determine if the therapy should be made widely available. Only then would government and private insurers be likely to pay for them.

Access to gene therapy, like genetic information services, will depend on the nature and amount of benefit that it would provide, particularly in comparison with non-gene-therapy alternatives. Even if it promised enormous benefits, however, gene therapy likely will not be covered by insurance if the cost is excessive.

Gene therapy may be expensive because of a number of factors. It may be labor-intensive, requiring repeated hospitalizations or the services of a large health care team. It may employ new drugs or medical devices that are protected from price competition by patent monopolies. In the case of gene therapy for rare disorders, the manufacturers may increase the price of the treatments to cover high per-unit research and development costs. Further, the government or powerful third-party payers, pursuing a "centers of excellence" approach that restricts the performance of certain medical services to high-volume institutions, may reduce the supply of the services, thereby raising prices. Finally, there may be only a few specialists who are able to provide the therapy, allowing them to command high fees.

Like many new technologies, it might be thought that even if gene therapy were expensive when it was first developed, the price would come down over time so that eventually it would be affordable. However, gene therapy might still be too expensive, even if its per-unit price declined following its introduction, if the demand for it was large. Even though the price of outpatient kidney dialysis declined from about $138 per session in 1974 to about $54 per session in 1989 (in 1974 dollars),[81] total government spending for dialysis increased from $229 million in 1974[82] to $5.4 billion in 1988[83] because it was being used in ever-increasing numbers of people. Similarly, even if an individual gene therapy were relatively inexpensive, the total number of therapies could be so large that collectively these therapies would have a significant impact on total health care spending.

Apart from a consideration of costs and benefits, access to gene therapy is likely to be limited because of religious and moral objections. We identified in Chapter Three the controversy over germ cell therapy. An even more contentious subject, at least at present, is abortion. If political, moral, and religious concerns continue to restrict abortion access, poor people who lack health insurance may be left without any genetic interventions. It might seem odd to mention abortion in a discussion of gene therapy. But abortion is one method of preventing an individual from suffering from a genetic illness or defect. In fact, given the expense of what might come to be considered more conventional gene therapies, abortion is likely to be the primary "preventive" technique, other than abstinence or contraception, that will be available to persons of limited means.

Another impediment to the accessibility of gene therapy is the refusal of most health insurers to pay for *in vitro* fertilization. As mentioned in Chapter Three, for the foreseeable future, the main techniques for directly manipulating a person's genetic endowment are likely to be aimed at embryos that are fertilized in the laboratory, genetically altered, and then implanted in the uterus or allowed to come to term in an artificial womb. Just as genetic information services are "gatekeeper" technologies, in that individuals will not know to seek preventive or therapeutic treatments unless they become aware that they have a genetic disorder or predisposition, limitations on access to *in vitro* fertilization, at least for the foreseeable future, will cut off access to gene therapy that must be performed in the earliest stages of fetal development in order to be successful.

In vitro fertilization is important not only as a gatekeeper technology, but as a terminal technology. If a fertilized embryo is tested and found to possess a genetic disorder, the embryo can be rejected for implantation in favor of one that is found to be free of the disorder. If *in vitro* fertilization is not available, however, individuals will be unable to make this selection. Prospective parents would only have two alternatives. One would be for them to make the decision not to conceive, based on the results of genetic screening tests on them individually. The second alternative would be to wait until a child is conceived, test it in the womb, and then abort it if it were found to have inherited the undesired trait. Both of these approaches have been used successfully in some Jewish communities virtually to eliminate the birth of children with Tay–Sachs disease, which invariably kills those afflicted with it within the first few years of life. [84]

Coverage of Genetic Enhancement

The final category of genetic technologies is genetic enhancement. As described in Chapter Four, genetic enhancements, while not likely to be available until farther in the future, are nevertheless likely to be exceptionally valuable to those who obtain them. It, therefore, bears discussing whether or not they will be widely accessible.

In contrast to the earlier discussion of genetic information services and gene therapy, the answer in the case of genetic enhancement is readily discernible. The closest existing analogy to genetic enhancement is cosmetic surgery. Insurers only pay for cosmetic surgery to correct a congenital anomaly or a disfigurement caused by accidental injury or disease. Cosmetic surgery for the purpose of improving appearance that is within normal limits is not covered. If insurers continue their present policy, they also will refuse to pay for genetic enhancement.

A recent court case illustrates how far insurers will go in order to avoid paying for cosmetic surgery.[85] In this instance, the insurer was the Idaho Medicaid program. The patient was an eight-year-old boy who had been diagnosed with Attention Deficit Disorder and diminished mental capacity (he was still in first grade). He qualified for Medicaid due to his disabilities. He also was diagnosed with a congenital deformity of his ears: they were grotesquely large. A psychiatrist determined that the boy would suffer damage to his already fragile self-esteem if he were teased about his ears. The boy's physician recommended otoplasty, a surgical procedure to reduce the disfigurement. But the Idaho Medicaid program refused to pay for the operation. The state took the position that the surgery was not "medically necessary" and, therefore, was not covered under the Medicaid law because it would not affect his hearing and would not "stabilize" or "improve the function" of the boy's ears. Under the state's interpretation, emotional, mental, and psychological factors were not recognized in considering functional improvement. The Idaho Supreme Court upheld the state's decision.

If third-party payers treat genetic enhancements the same way that they treat cosmetic surgery, then the only people who will obtain access to them are those who are able to afford to purchase them with personal funds. The more expensive the enhancement technologies, the more limited this group may be.

ACCESS TO GENETIC TECHNOLOGIES BASED ON ABILITY TO PAY

We saw before that there are a number of approaches to allocating access to medical technologies that are in short supply. Some focus on the characteristics of the patients seeking access, such as social worth, while others focus on the characteristics of the technologies themselves, such as a comparison of their costs and benefits. But virtually all of these approaches have one factor in common: whatever other basis they purport to use to allocate access, they also allocate access on the basis of the patient's wealth. This follows from the fact that patients who can afford the expenditure can simply buy the service they desire even if it is not covered by their public or private insurance plan.

Wealth-based access has long been part of our health care system. As mentioned in Chapter Four, during the dialysis crisis in the 1960s, patients who could afford the approximately three thousand-dollar cost of a dialysis machine, plus the service charge of as much as three hundred dollars per treatment, could obtain dialysis. If necessary, patients paid their physician to purchase a machine for their personal use. [86]

Wealth-based allocation is still very much a part of our current health care system. The clearest example is transplant organs. There are a host of medical criteria that patients must meet in order to be eligible to receive an organ transplant. But patients also must be able to demonstrate that they can pay for the procedure—either through insurance or with personal funds. [87] If patients' insurance policies do not cover transplants, patients have to come up with the funds themselves before they will be placed on a waiting list.

It might be thought that poor patients can still get transplants because they will be covered under Medicaid. But as the Oregon experience demonstrates, many poor patients do not qualify for Medicaid, either because they do not fit into one of the eligible patient categories or because their incomes exceed the state income threshold. Even if a patient qualifies for Medicaid, he or she still may not receive a transplant, since states have the discretion to refuse to provide coverage of organ transplants other than kidneys under their Medicaid programs. [88] A survey of state Medicaid programs we conducted showed that many states, in fact, do deny coverage.

Wealth-based access to health care also is reflected in the Oregon Medicaid approach. Wealthier Oregonians will still be able to purchase services that are not ranked high enough in the priority list to be subsidized by the government for lower income individuals. Indeed, the Oregon priority system was specifically designed to apply primarily to those residents whose incomes are below the federal poverty level and who cannot afford to purchase medical care with their own funds. These people have no way to obtain coverage for services that are below the priority cut-off point.

President Clinton's health care reform plan also contemplated that persons with greater resources would be able to purchase medical services that were not included in the basic benefits package. One way they could have done this would have been to purchase supplemental insurance policies that covered additional, nonbasic services. Or, they simply could have bought the services with their personal funds.

We conclude this chapter, then, with a key observation. We have been discussing whether people will obtain access to genetic technologies. We have seen that the answer depends on three factors: whether there is a supply shortage created by technical conditions; whether the technologies are covered by public or private insurance; and whether people have the information they need to seek access. But even if the supply is limited, even if insurance does not cover the service in question, and even if information is hard to come by, these impediments may not stand in the way of persons who can purchase the genetic technologies with their own resources. They do not need insurance. They are able to buy the information they need or hire experts to obtain it for them. Even in the case of supply shortages, they may be able to purchase the technology given enough funds, as they did with dialysis machines in the sixties. Alternatively, money may be one of the prerequisites to obtain access under a distribution system such as the one society uses to allocate organ transplants.

The likelihood that genetic technologies will be available to some people but not to others, and that a major determinant of access will be wealth, raises profound social issues to which we next turn.

6

Genetic Technologies and the Challenge to Equality

As it stands now, many future genetic technologies will be accessible only to those people who have insurance coverage or who can afford to purchase such technologies with private funds. This will deny access to persons who lack health insurance—currently estimated to be around forty million individuals—unless they have sufficient personal wealth. Even those persons who possessed health insurance might be unable to obtain access to expensive genetic technologies that were not part of the insurers' benefits packages, either because the technologies were experimental or because they were too costly. One particular set of genetic technologies—genetic enhancements—are likely to be excluded entirely from coverage and will only be available to those persons who can purchase them with private funds.

Some people, particularly some economists, will defend this method of allocating genetic services. Allocating resources on the basis of willingness to pay is consistent with the belief that, all other things being equal, the best way to identify and measure the strength of an individual's demand for goods and services is to rely on the individual's own choices. This enables the market to price goods and services efficiently and to avoid wasting production capacity by producing goods and services for which people are not willing to pay. Allocating resources on the basis of a person's willingness to pay arguably also maximizes the individual's decision-making autonomy. No one else—not the government or a private insurance plan—is in a position to interfere with the individual's wants.

These attractions account for the popularity of proposals to replace government health insurance programs with medical savings accounts. These would allow people to put earnings in interest-bearing accounts without paying taxes on them and to withdraw funds to purchase desired medical care. Money in the account that was not spent could be used for other purposes or become part of a person's estate upon death.

But willingness to pay runs into a fundamental problem. It might be a satisfactory and perhaps even superior way of allocating access to genetic technologies so long as individuals are able to afford the technologies they desire. Part of what economists who advocate willingness-to-pay approaches mean when they assume that "all other things are equal" is that everyone has roughly the same amount of wealth. But as we saw in Chapter Five, and as we know from personal experience, this clearly is not the case. Not only are some people wealthier than others, but some have health insurance while others do not, and some insurance plans cover services that others exclude.

Another assumption that economists make is that everyone has the same basic needs for the desired goods, for example, the same risk factors that cause them to need genetic technologies. But this assumption also is false. People differ in their inherited characteristics and lifetime experiences so that some need different types and amounts of health care than others. A willingness-to-pay approach could still result in an efficient allocation of genetic services if those people who had a greater need for these services also were more likely to possess the wealth to pay for them. But this, too, is not the case. As Richard Epstein, a distinguished law professor who is a proponent of free market economics, points out, ability to pay does not correlate well with medical need.[1] Sick people are not wealthier than healthy people. If anything, the opposite is true; poor people tend to suffer more illness than those with greater financial resources, due to the lack of access to timely care and to the poorer quality of their nutrition and living environments. Under a pure willingness-to-pay approach, then, some people with a need for genetic technologies would have access to them while others with the same need would not. From an economist's standpoint, this result would be inefficient.

Economic efficiency is not the only reason to be concerned about a system that would make many important genetic technologies accessible only to those who could pay for them. Arguably, nothing is as essential to a good life as good health. If we think of some desired goods as "wants" and others as "needs," good health is as close as anything to being a "need." To the extent that health care is necessary for the good life, it is a "primary" good, along with such necessities as food and shelter. Even if we thought that people ought to obtain access to most goods on the basis of what they chose and what they could pay, we might believe that we had a collective responsibility to ensure that everyone had at least a minimum set of primary goods, including access to health care.

If health care in general is a primary good, so are many genetic technologies. As described in Chapter Three, people who cannot obtain genetic services will be denied critical health benefits. Without access to genetic testing, they will lack much of the ability to forestall or reduce the severity of genetic disorders. People lacking access to effective gene therapies will be forced to rely on less effective conventional therapies that, in many cases, somewhat ameliorate the symptoms of their ailments rather than curing them. The more the Human Genome Project leads to the discovery of genetic tests and successful therapies for genetic disorders, the more the health of those without access to these services will decline in comparison to those with access.

Genetic testing and therapy are not the only types of genetic services that might be considered primary goods, however. As discussed in Chapter Three, genetic manipulation may be able to alter significantly a person's physical appearance—such as his or her height or attractiveness. Genetically engineered human growth hormone may make it possible to increase height, although it is unclear whether this will occur without troublesome side effects.[2] Once the right genetic precursors are located by the Human Genome Project, the same techniques that permit us to correct genetic errors should enable us to alter other physical attributes such as strength, stamina, and visual and aural acuity.

Ultimately, the most significant result of the Human Genome Project may be the ability to enhance a person's mental abilities. We are only beginning to understand the extent to which mental characteristics are the product of a person's genetic endowment. Research in this area is mired in controversy, with critics concerned that it could lead to everything from cuts in social welfare programs and increased discrimination against certain genetically defined races and ethnic groups to Nazi-like eugenics programs in which governments try to breed a better citizenry.[3] Even so, it may only be a matter of time before we are able to improve a person's memory, concentration, or IQ through genetic manipulation.

While lack of access to gene testing and gene therapy would deprive people of primary health care services, lack of access to genetic enhancements could deny them the ability to satisfy primary social needs. Many people would consider the resulting society to be intolerably unjust. One of the leading experts on justice and health care, Norman Daniels, maintains that individuals must have a fairly equal opportunity to obtain those health care services that will provide a "normal range of opportunity."[4] Daniels' notion of a normal opportunity range refers to those opportunities he considers necessary for an individual to formulate and carry out a life plan. Health care is important in this scheme because a certain level of healthy physical and mental functioning is necessary for an individual's prospects to fall within an acceptable range of opportunity. If physical health conditions and mental abilities achievable with genetic technologies come to be regarded as necessary to enable individuals to carry out a life plan with a normal range of opportunity, then it would be unjust to deny individuals that chance.

As discussed in Chapter Three, many individuals would not obtain access to a number of important genetic technologies under the present system. Access would be unequal, and would depend primarily on whether an individual had insurance that covered the genetic services in question, or could pay for them with private funds.

Philosophers and political theorists have struggled for centuries to construct models of society that justify certain types of inequality. One way of analyzing whether our prediction of unequal access to genetic technologies would be acceptable from an ethical standpoint is to see if it can be justified under any of these theories.

A leading theory of inequality is utilitarianism, developed originally by Jeremy Bentham and John Stuart Mill in the nineteenth century. The principle of utility is well-known. It holds that we should act (or arrange our institutions) so that the results of the act bring about the greatest good for the largest number of people. The "good" may be a general good, such as happiness, or a specific good, such as health care.

Utilitarians speak of maximizing either aggregate utility or average utility. Distributing a good to maximize aggregate utility refers to adding together the utility for each individual affected.[5] If we conceived of utility in terms of units, for example, one would simply add together the units that all persons affected would receive by a particular distribution. For example, if there were five individuals, *A* through *E*, and *A* received three units, *B* received five units, *C* seven, *D* four, and *E* six, the total utility gained, or aggregate utility, would be twenty-five units.

Distributing a good to maximize average utility refers to the aggregate utility divided by the number of persons who receive it.[6] Thus, the average utility in the example above is five (twenty-five units divided by five persons affected).

The principle of utility might be applied to the allocation of health care resources by assessing the consequences of alternative distribution systems and determining which system would provide the greatest number of units of good (either in the aggregate or on average). One way of evaluating the consequences is through the use of cost-benefit analysis. The benefits and costs of alternative distributions could be calculated to determine the distribution with the greatest amount of net benefit or the best ratio of costs to benefits. A system that delivered five units of benefit at three units of cost would be preferred over a system that delivered five units of benefit at four units of cost.

Preventive health services often fare well under a utilitarian scheme because their cost is low and the benefit is generally deemed to be high, in part because often they prevent serious illnesses and their benefit endures for a long time. Utilitarians, therefore, might call for widespread access to genetic testing and gene therapy that detect and prevent serious genetic illnesses.

"Rescue medicine" is another story. The cost of providing gene therapy to save *one person* with a catastrophic illness may be the same as providing genetic testing and preventive therapy for *many individuals*. The total units of utility produced in both cases might be the same. A genetic test that would lead to preventive treatment might add ten years of life to five individuals, while a life-saving gene therapy might save the life of a child with a life expectancy of fifty years. In each case, the same amount of utility, measured in terms of additional years of life, would result, and a utilitarian would have trouble deciding whether to provide access to one or the other. The fact that the benefit was enjoyed by one individual in the first case and by five individuals in the second would be irrelevant. Some would say that utilitarianism produces unjust results on this account.

Another problem with utilitarianism concerns how to determine just what the distributive result would be. In terms of health care, just as there are many different ways of measuring costs and benefits, so there are many different ways of distributing access to health care that would be consistent with utilitarian principles. Yet each of these approaches would yield a different result.[7] To maximize the number of years of life that would be saved by a genetic therapy, we might limit access for the elderly. If benefit to society were the primary objective, we might limit access for criminals and give preferences to philanthropists and medical researchers.

In terms of genetic technologies, utilitarian theories would support a wide range of distributive systems, depending on how utility, costs, and benefits were calculated. Utilitarians might sanction a broad pattern of distribution in which many people, or at least a majority, obtained some access to some services. They might opt instead for a highly restricted system in which access to genetic technologies was concentrated in those individuals with the poorest genetic endowments. Or, utilitarians might support giving access only to those who already possessed good health and superior traits, on the theory that the resulting super-persons would provide the most benefit to society, thereby indirectly maximizing everyone's utilities.

There simply is no ready means of determining which genetic technologies would be available to which persons under a utilitarian approach. The only point that is obvious is that the interests of individuals can be subordinated to the interests of others in order to maximize the sum of utility. Utilitarianism, therefore, could sanction a broad range of inequality in terms of access to genetic technologies.

A number of other theories of justice insist that individuals have rights that cannot be sacrificed in order to create more utility for others. The most well-known and influential of these rights-based theorists is John Rawls. In his landmark work, *A Theory of Justice*,[8] Rawls asks us to imagine that we are in a position to choose principles to govern societal institutions without knowing the kind of society in which we would live or the position in society into which we would be born. Under these circumstances, argues Rawls, we would choose two principles.

One is the principle of greatest equal liberty. According to this principle, "each person is to have an equal right to the most extensive basic liberty compatible with a similar liberty for others."[9] In other words, we must maximize liberty, but we may do so only to the extent that it is possible to ensure the same liberty for everyone.

The second principle Rawls calls the "difference" principle. It holds that people should have equal opportunity to participate in important societal institutions and that the only differences in wealth or social status that would be acceptable would be those that would make everyone better off. In other words, one individual or group may have an unequal advantage only if all others realize some betterment from the unequal distribution of the wealth or benefit.

Although Rawls himself has not written about the implications of his theory of justice for health care, others have tried to derive working policies using his principles of justice.[10] Each attempt has encountered difficulties in taking what is a very abstract theory of distributive justice and trying to fashion particular rules to govern hard choices in actual situations. Much of the problem stems from what others see as one of Rawls' strengths: the flexibility of his principles. By being flexible, his principles allow for any particular society to set up institutions in a number of different ways, depending on the culture, the scarcity of resources, the peculiarities of the underlying distribution of wealth in that society, and so on. Yet this seems to permit a wide range of choices for specific policies. For example, Rawls' second principle, encompassing both equality of opportunity and the difference principle, has been invoked to support both the most minimal rights to health care and the most extensive access that society can afford. In terms of genetic technologies, Rawls might be comfortable with a considerable degree of unequal access, so long as he could be persuaded that this was the approach in which everyone had the greatest liberty and in which everyone derived at least some benefit from the inequality of access.

Another theory, libertarianism, rejects the notion that society can compel individuals to relinquish their wealth or power for the benefit of others. According to Robert Nozick, a leading libertarian philosopher,[11] people are entitled to the holdings they acquire justly. Answering the question of whether a holding was justly acquired requires looking to the past. Does the person have a natural talent or ability? A libertarian would say that income or wealth acquired as a result of a natural talent is justly held and deserved in that sense. If the wealth is justly held, it may be voluntarily exchanged for other goods.[12] Thus, if a person has perfect pitch and becomes a famous and wealthy opera singer, the holdings she has as a result are justly held and she may exchange them for other goods and services, e.g., health care. According to this view, requiring the opera singer to give up some of her wealth so that others might also receive health care is unjust. Rather, a libertarian would rely on people making charitable gifts for this purpose.

Philosopher Paul Menzel has argued that the libertarian view can be used to design an ethically sound model for unequal access to health care in a democratic society.[13] Menzel looks at individuals' willingness to pay to avoid risks of future harm (pain, suffering, disability, or early death from illness). A good example is health insurance pools in which groups of persons choose a plan to protect themselves against future risk of illness. The key to Menzel's approach is that he holds individuals to their past preferences. A person who purchased a cheap insurance policy would be deemed to give consent to forgo services not included in the plan, including services needed to treat a serious or life-threatening illness. It may seem unfortunate both that the particular illness has struck the individual and that its treatment is not covered by the plan that the individual purchased in the past, but, according to Menzel, it would not be unfair to withhold the treatment the individual now needs and wants.

At the same time, if the individual has the resources to purchase the services, the libertarian would argue that she must be allowed to do so. A person who needed an autologous bone marrow transplant for breast cancer that was not covered by her insurance plan should be able to purchase those services in the free market.

The libertarian viewpoint thus is clear about how it would allocate access to genetic technologies. Those who could afford to purchase them would receive them, and except for those who received services through charity, the rest of the population would not. The result would be a much greater degree of inequality than would result under our current system, where government programs, such as Medicare and Medicaid, and the tax deductibility of insurance premiums, subsidize health care.

A major objection to libertarian theory generally, and to Menzel's version of it in particular, is that it rests on the assumptions that individual holdings have been justly acquired and that we can distinguish those that have been so acquired from those that have not. Can we reasonably say that anyone's wealth is the result of an unbroken chain of just acquisition and transfers? If not, then perhaps some redistribution is warranted to rectify past injustices.

Unlike utilitarian and rights theories, the libertarian view also does not take into account the overall benefit of its citizens' good health to a society.[14] A healthy citizenry contributes to the general level of production and flourishing of a society; an unhealthy citizenry impedes progress. Therefore, there might be value even in a libertarian society in funding public institutions that promote good health.

One particular set of theories, called communitarianism, goes even further and emphasizes that the common good, or the good of society, takes precedence over the good of the individual. Daniel Callahan has written extensively on this topic as it relates to health care.[15] He claims that the individual only benefits when the community in which he or she lives is good (however good is defined). The primary effort should be to assure societal welfare; individuals then receive the benefit of living in a just society. Similarly, the provision of health care services should be geared toward improving or maintaining the overall health of a society, rather than ensuring that each individual's needs or wants are met. Citizens should be guaranteed a decent baseline of health care,[16] but everyone should not get whatever they need or want. Where curative treatment is not provided, an individual must receive supportive care. According to Callahan, caring is the foundation of respect for human dignity and worth and must always be provided in a just society.[17]

In regard to distributing genetic technologies, a communitarian approach would be willing to constrain individual access to the extent necessary to achieve communitarian objectives. Thus, if it were felt that certain technologies, such as germ cell engineering, were a serious threat to the safety or stability of society, communitarians would support restrictions on the personal freedom to acquire them. On the other hand, if allowing some individuals to obtain access to certain genetic technologies, even though others could not, were believed to provide sufficient benefit to the community, there would be no communitarian objection.

Identifying and achieving communitarian objectives, however, would be difficult, and perhaps highly controversial. Would it be appropriate to require certain persons to relinquish access to life-saving gene therapy in order that the community might devote the resources to others? What about the elderly? Could we say that they "owed it" to the community to decline expensive genetic treatments so that younger people might live, or so that genetically healthier babies might be born?

Daniel Callahan has advocated just such a position. The best method of allocating access to medical technologies, in his opinion, is according to age. To cope with the scarcity of medical resources, he originally proposed that scarce medical resources be concentrated on the population below a certain age cut-off.[18] Although he never specified the precise age limit, it seemed to be around eighty years old. People over the age limit would be given access only to comfort care, not to what Callahan considered to be heroic, life-saving treatments, such as dialysis or major organ transplantation.

Callahan's proposal for age-based rationing triggered a storm of protest. Advocates for older Americans objected to singling out age as the chief characteristic that determined whether a patient would be given access to particular treatments.[19] Others criticized Callahan's approach as really a wealth-based rationing scheme in disguise, since the only elderly patients who would be denied medical services by his scheme would be those who could not afford to purchase them with their own funds. Callahan attempted to counter this by suggesting that the government should prevent technologies for the extremely elderly from being developed in the first place, but he ran into insurmountable problems in defining which technologies the government should restrict, since virtually all technologies used in elderly patients also are employed in the treatment of younger persons whom Callahan thought were entitled to those treatments.[20]

Callahan has since switched his approach to one that does not emphasize prioritization based on age so much as on an assessment of the costs and benefits of medical technologies.[21] Moreover, he is less focused on mandatory government limits on access than on convincing people to limit voluntarily their consumption of expensive health care services, especially when these services merely prolong rather than improve the quality of life. But it is still clear that his goal is to persuade individuals to sacrifice their personal interests for the greater good of the community.

In the end, it is evident that all of these theories would tolerate a significant amount of unequal access to genetic technologies. Utilitarians and rights-based theorists such as Rawls would accept inequality on the grounds that it maximized total utility or benefitted the least well-off along with others, although Rawls might have trouble with a distribution method that limited liberty. Libertarians would sanction a free market system, or one in which individuals were held to their earlier decisions about how much to spend on health care; in either case, inequality would be rampant. Since communitarians would sacrifice the welfare of the individual for the welfare of the community, they would support an unequal system if they thought it necessary in order to promote community goals, such as containing health care spending. If we are worried about unequal access to genetic technologies, in short, we need to look beyond these theories for the reasons for our concerns.

Let us suppose, therefore, that we proceeded to allocate access to genetic technologies according to the principles that govern the current system. The result, it will be recalled from Chapter Five, would be that many technologies would be available only to those with insurance or to those who could pay. Moreover, one whole set of technologies—genetic enhancements—would not be covered by insurance at all and would be accessible only to those individuals with the personal wealth to purchase them. What would such a society be like?

In this "postgenorevolutionary" society, those who could afford to would have their children by *in vitro* fertilization, which, as now, would not be covered by insurance. They would employ the techniques described in Chapter Three to select only superior embryos for implantation, or would manipulate the genetic characteristics of the embryos to make them free of inherited diseases. Most people would be relegated to whatever gene therapies government or private insurers chose to cover. Even these technologies would be beyond the reach of persons who could not afford to pay the corresponding deductibles and copayments. Insurers would most likely cover only relatively inexpensive therapies that were highly effective.

The one type of genetic technology that would be most likely to be widely available, even for those persons who were not insured, would be genetic screening services. Insurers and government programs would provide generous coverage of these services in the hopes that people would avail themselves of the most common techniques of gene therapy: birth control and abortion. In the face of the growing availability of genetic tests for inherited diseases, the government would come under increasing pressure not only to reject pro-life opposition to abortion, but to cover abortion generously under government health insurance programs and to provide inducements, such as tax advantages, to persons who would agree to be screened and then decline to give birth to genetically diseased children.

Genetic enhancements, meanwhile, would be available only to a narrow, wealthy segment of society. This "genobility," already privileged by its wealth, would experience an unprecedented burst of positive evolution. The privileged status of its members would become more and more unassailable, particularly if genetic enhancements were installed through germ cell manipulation and, therefore, were passed on from one generation to the next.

Widespread genetic screening, coupled with birth control and abortion, whether voluntary or under pressure, would enable people to avoid passing on genetic disorders to their offspring. But there would be little opportunity for the average person to ascend the genetic ladder. There might be some intermarriage between the genetic aristocracy and the genetic underclass, reminiscent of the "poor boy (or girl) marries rich girl (or boy)" scenario, but this is not likely to be very common. Studies show that, by and large, Americans tend to marry people very much like themselves—that is, from their own social class. [22] Occasionally an Horatio Alger story would enable a member of the genetic underclass to accumulate enough wealth to purchase the genetic means to become part of the aristocracy, but this is likely to occur even less frequently than it does now, as genetically engineered advantages enable the "genobility" to monopolize the most lucrative jobs and investment opportunities.

In short, providing access to genetic technologies according to current coverage policies would create a widening gulf between the genetically privileged and the genetic underclass. One group would be virtually free of inherited disorders, would receive powerful genetic therapies for acquired diseases, and would be engineered with superior physical and mental abilities. The other group would continue to suffer from genetic illnesses and would have to content itself with less effective, conventional medical treatments. Its members would be able to improve their mental and physical traits only through comparatively laborious traditional methods of self-improvement.

The division of society into a genetic aristocracy and a genetic underclass would have momentous consequences not only for individuals, but for democratic society as well. It would undermine the fundamental precept upon which such a society rests: the maxim of social equality. As the Declaration of Independence states: "We hold these truths to be self-evident: that all men are created equal; that they are endowed by their creator with certain inalienable rights; that among these rights are life, liberty, and the pursuit of happiness." These are more than just lofty, idealistic sentiments. They are the glue that holds our society together.

Social inequality is inherently destabilizing. As one sociologist has observed:

> Inequality in the distribution of rewards is always a potential source of political and social instability. Because upper, relatively advantaged strata are generally fewer in number than disadvantaged lower strata, the former are faced with crucial problems of social control over the latter. One way of approaching this issue is to ask not why the disprivileged often rebel against the privileged but why they do not rebel more often than they do.[23]

Obviously, people are not endowed with the same genetic traits. The Human Genome Project will have taught us this at least: that some people are healthier, prettier, craftier, stronger, or more intelligent than others, due in large part to their genetic makeup. But these differences do not make it impossible to ground society on a foundation of social equality. Western democratic societies accommodate these differences by means of a widespread belief in equality of opportunity. What matters is not that people are the same, or even that they believe that they are equal. What matters is that people believe that they have just as good an opportunity to succeed as the next person:

> Whereas most Americans are willing to tolerate sizeable inequalities in the distribution of resources, they typically insist that individuals from all backgrounds should have an equal opportunity to secure these resources.[24]

The belief in equality of opportunity performs the key function of enabling our society to remain relatively stable politically despite significant actual inequalities. As a noted political theorist, John Shaar, observes: "No policy formula is better designed to fortify the dominant institutions, values, and ends of the American social order than the formula of equality of opportunity, for it offers *everyone* a fair and equal chance to find a place within that order."[25] The lower classes accept the existing order because a sufficient number of them believe that there is equality of opportunity. In essence, equality of opportunity means that it is possible for a person, through talent and effort, to move up the social ladder.

The importance of upward mobility was recognized as far back as the ancient Greeks. In *The Republic*, for example, Plato conceptualized that society was divided into groups of "gold," "silver," and "bronze" individuals. Individuals were born into these conditions, he acknowledged, but they had to be allowed to move up or down, depending upon their abilities. Thus, bronze children of gold parents and gold children of bronze parents deserved to be given their rightful places in the hierarchy of leadership.[26]

Upward mobility provides the main evidence that equality of opportunity, in fact, exists. A belief in upward mobility also motivates members of the lower classes to produce and achieve in the hopes that they will move up on the social scale. More importantly, upward mobility is a primary stabilizing mechanism in unequal Western democratic societies:

> There is a long-standing and widespread idea that circulation of individuals and families among the different levels of an otherwise divided society acts as a kind of safety valve to keep the pressures of discontent low. Hopes among the unprivileged may then, it is assumed, be centered on personal achievement rather than on collective resistance, rebellion or revolution.[27]

Marxist scholars, for example, attribute the lack of revolutionary fervor among lower classes in Western societies to their belief in upward mobility:

> Capitalism is taken to lose a good part of its sting in so far as wage-earning dependency is not a fixed and inherited condition. Class barriers are seen as dissolving, the more individuals can move across the face of the social structure; as the possibility of movement helps to generate personal aspirations among workers that either they, or at least their children, may be able to reach the security of supervisory, managerial or professional positions.[28]

As one famous sociologist remarked almost seventy years ago about the impact of upward mobility on the lower classes: "Instead of becoming leaders of a revolution, they are turned into protectors of social order."[29]

Upward mobility not only soothes the restlessness that inequality creates among the lower classes, but syphons off those individuals who are most accomplished and, therefore, most likely to be the instigators of unrest, reassigning them to higher status positions: "Mobility provides an escape route for large numbers of the most able and ambitious members of the underclass, thereby easing some of the tensions generated by inequality."[30] Even if individuals come to realize that they are destined to remain at a fixed social level or, even worse, to move down, equality of opportunity can continue to tranquilize them by holding out the possibility that the pattern will be reversed in their children.[31]

A genetically stratified society such as we envision would challenge the concept of social equality in three fundamental ways. By enabling a genetic aristocracy to achieve greater genetic health and talent than those who do not have access to genetic technologies, genetic stratification would substantially increase actual inequality. By allowing the members of this aristocracy to manipulate their genetic endowment and possibly even to pass genetic advantages on to succeeding generations, it would undermine the belief in equality of opportunity. Finally, stratification would freeze up the crucial safety valve of upward social mobility. The genobility would monopolize desirable occupations and fill higher status social roles. Members of the lower classes no longer would be able to count on traditional methods of advancement, such as education and intermarriage, to improve their status.

If the reduction in upward mobility were substantial enough, the lower classes, who could not afford to better themselves genetically, would remain locked into their genetic class. With genetic superiority predisposing individuals to social and economic success, and this success in turn permitting families to preserve their genetic advantages, membership in the upper classes would be genetically preordained by virtue of the genetic opportunities given to offspring by their parents. Membership in different genetic classes would become a matter of inheritance.

Inherited, largely fixed social status is no stranger to the human condition. In feudal Europe, individuals were born into their classes and, with rare exceptions in which peasants were able to obtain education in religious institutions or became apprenticed and eventually squired to knights, they stayed there. In slave-owning societies, people were born into bondage, although they could be freed at the pleasure of their masters. Perhaps the most immutable type of social strata is represented by caste systems, such as those of traditional India.

Static social strata may be compatible with some forms of human society, but they are not compatible with that type of society known as Western democracy. The demise of feudalism, slavery, and the caste system is generally believed to have been indispensable to the rise and survival of democratic systems.

Genetic social stratification, then, clearly threatens democracy, but it is not clear how seriously. Perhaps society will adapt to the social artifacts of the genetic revolution and the result, while markedly different from present arrangements, will be relatively stable. The planetary motto of Aldous Huxley's *Brave New World*, in which scientists produced large numbers of standardized individuals, was "Community, Identity, Stability." As one of Huxley's characters observed: "You really know where you are."

In one vision of our future, for example, the genetic underclass might cede power to their genetic superiors in return for enjoying the material benefits made possible by genetic advances. The underclass would accept the division between social strata, and be content with being upwardly mobile only within the confines of their own class. The genobility, in turn, would rule according to enlightened principles of noblesse oblige, being careful to permit sufficient benefits to trickle down so that political and social equilibrium was maintained. A democracy of sorts might even persist, with the underclass electing representatives who either belonged to the upper class or who were committed to preserving its privileges. Such a system might not look very different from our own, given the extent to which we increasingly elect representatives who are considerably more privileged than their constituents.

Such a system seems highly unstable, however. For one thing, the members of the upper class would need great self-control to avoid overreaching. At a minimum, they would need to maintain effective means of monitoring and regulating the behavior of their peers to prevent antisocial excesses of greed. Even so, the media—assuming it remained free—could be counted upon to glamorize the lifestyles of the genetically rich and famous in graphic contrast to the more mundane existences of the underclass. The fact that genetically superior persons owed their advantages to accidents of birth or to their starting positions of wealth would reduce the admiration that the underclass might feel for them, while leaving unchanged, or perhaps increasing, underclass admiration for those who had gained their privileged status through hard work or the development of great skill.

Such a system would be vulnerable to demagogues who came to power by promising to redistribute genetic endowments more evenly. Assuming that the principle of one-person/one-vote persisted, a numerically inferior genetic upper class could be out-voted by the underclass. Congress could become dominated by elected officials pledged to employ the full force of government to rectify genetic imbalances. (A number of these steps are discussed in Chapter Seven.)

The genobility would respond with reprisals in an effort to preserve its privileged status. These could range from threats to withhold the fruits of genetic medicine from nonprivileged segments of society to overt interference with the democratic process. At the least, the genetic upper class is liable to have amassed sufficient wealth and influence to enable it to control the media, which would permit it to affect the outcome of elections in a manner quite out of proportion to its numbers. Efforts by the underclass to preserve its majoritarian hegemony may prove no more successful than have current efforts to reform campaign financing in order to dilute the power of special interests.

The end result might be an era of social chaos as the system swung in ever-widening arcs between rule by underclass demagogues and domination by the genetic aristocracy. This could degenerate into mob rule and anarchy. To rid itself of its status as the class of the genetically disadvantaged, the mob might even destroy the scientific foundations of the genetic revolution, perhaps by physically dismantling research centers and erasing mapping and sequencing data.

From our current vantage point, the exact nature of postgenorevolutionary society is obscure. Alternatively, postgenorevolutionary society could devolve into totalitarian rule by a genetic autocracy. The genetic upper class would employ whatever repressive techniques were needed in order to obtain power and keep the underclass in check. Given the advances in genetic science that would have made genetic class distinctions so marked, techniques might even be developed to manipulate the underclass genetically to make it more docile.

What is clear is that the genetic technologies of the future come with a curse. They promise great advances in our ability to forecast and forestall disease and to improve the capabilities of the human species. But, judging from the increasing scarcity of resources and the dislocations that are likely to result from distributing genetic technologies according to current allocation principles, genetic technologies represent a serious and fundamental threat to our social and political system. Rather than blessing us with unprecedented social progress, the genetic revolution may plunge us into a new Dark Age.

But perhaps we are going too far. This bleak future arguably is only one of a number of possible alternatives. Many people may dismiss our warning that the development of genetic technologies could provoke the decline of democratic civilization. This is nothing more, they may say, than the equivalent of Chicken Little alarming the barnyard due to a genetic acorn.

The introduction of these technologies will be so distant and so gradual, some may chide, that society will have a chance to adjust to them naturally without cataclysm. Given enough time, the means of coping with these advances will evolve within our democratic institutions, just as these institutions have adjusted to other profound technological changes, such as the advent of electronic media, nuclear weapons, and the computer. The postgenetic social and political system may look different, we may be told, but it will retain the tell-tale characteristics that make a relatively stable, free society possible.

The problem with this viewpoint is that it ignores the degree to which genetic advances, if distributed along current lines, will alter the fundamental assumptions that underlie democratic society. There has never been a challenge to the principle of equal opportunity as powerful as the threat posed by these technologies, with the possible, hardly democratic, exception of slavery.

Another more rosy view of the future rejects the assumption that access to genetic technologies will be confined only to those who can afford to pay for them. For one thing, genetic technologies may not be as costly as we predict. Genetic technologies may become so inexpensive, both in an absolute sense and in comparison with the costs of providing conventional medical care, that we will be able to generate enough savings to provide genetic services to everyone. But the fact remains that, when they are first introduced, genetic technologies, like most new medical services, are likely to be very expensive. Moreover, as noted earlier in Chapter Four, even if the price of a genetic technology decreases over time, the total cost is likely to increase as the usefulness of the technology spreads to broader numbers of patients.

A related viewpoint is that we may save so much money by preventing certain genetic disorders that we will have the resources to provide genetic technologies to everyone. This argument is a familiar one in terms of reducing health care spending. If we immunized enough people, or detected enough cancers early enough that they can still be treated, or stopped enough people from smoking, we are told, we would be able to redirect our health care expenditures so that we can provide such expensive services as hospitalization, surgery, and life-saving drugs to everyone who needs them.[32] Similarly, if we prevented people from dying from fatal or life-threatening genetic disorders, such as cystic fibrosis or inherited forms of cancer, we would save enough money so that we could provide genetic technologies to everyone.

The problem is that, surprising as it may sound, it simply is not clear that preventing illness saves money. No doubt it may reduce the costs of acute care in the short run, but in the long run, the people who would have died from acute ailments would go on to live longer and to contract the expensive, chronic illnesses of old age. For example, an article on the costs of cigarette smoking in the *Journal of the American Medical Association* found that, while every pack of cigarettes smoked increased health care costs by thirty-eight cents, it saved $1.82 in pension costs.[33] An earlier British study found that a 40 percent reduction in smoking would save that country sixteen million pounds in the first ten years, but would cost a net thirteen million pounds after thirty years due to the additional costs of social security payments.[34] Similarly, physicians studying prenatal care programs for low-income women have found no conclusive evidence that these programs save money.[35] This is not to say that we should not prevent disease when we can, but only that there may be better arguments for doing so than reducing long-term health care costs.

The final objection to a doomsday scenario is that, even if genetic technologies were tremendously expensive, and regardless of how much money we were already spending on health care, the government would do whatever was necessary to make genetic technologies available to everyone. As noted earlier in Chapter Four, this was essentially society's solution to the shortage of kidney dialysis machines in the late 1960s.[36] Congress included patients suffering from end-stage kidney disease within the Medicare program even if they were under the age of sixty-five or not otherwise eligible and mandated that dialysis was a covered service under Medicare. While a number of health policy analysts criticize this program because of its cost and because dialysis is not a cure and does not permit patients to enjoy a completely normal lifestyle,[37] others commend the program for saving lives and reflecting a societal commitment to the ill.

This same approach of including genetic technologies under Medicare, and extending eligibility to persons regardless of their age, could be employed in regard to genetic services. Or, in the event Congress enacted a national health care reform program that guaranteed universal coverage, it could include genetic services within the congressionally mandated basic benefits package.[38] If everyone received access to important genetic benefits, the adverse social effects of genetic stratification might be avoided. The only technologies that might be excluded from coverage due to cost would be those that were expected to yield only trivial benefits.

What technologies might be regarded as trivial? They might include genetic tests to detect disorders that could not be treated, or to identify extremely low probabilities that persons being tested would suffer from inherited disorders in the future, or that predicted the susceptibility to relatively minor ailments. Trivial genetic therapies might be those that treated or prevented minor disorders, or disorders in patients with no hope of long-term survival, or with severely compromised qualities of life, such as patients in persistent vegetative states or anencephalic newborns. In terms of genetic enhancements, perhaps those that altered relatively insignificant aspects of physical appearance, such as eye or hair color, might be denied.

However, this assumes that we could define which benefits were trivial. What might be trivial to one person might be important or even critical to another, or at least significant enough that, in order to get it, the individual would be willing to trade off something that others would regard as more desirable. At least, this is the basic assumption behind the well-entrenched ethical and legal doctrine of informed consent, which requires that a physician allow a patient to decide which medical interventions he or she will accept, based on the patient's own assessment of the relative importance of different sets of risks and benefits.[39] To avoid making blanket prioritization decisions that potentially could waste resources by providing certain genetic technologies to people who did not value them highly, we instead might have to establish limits on the amount that individuals could spend on genetic technologies—a sort of voucher system as described in Chapter Five—and then permit them to make their own selections.

The notion that important genetic technologies can be provided to everyone is unrealistic, however. It is naive to expect that the price of genetic technologies will drop to such an extent, or that the societal savings from their use will be so great, or that voters will be willing to commit sufficient funds, that we will be able to provide everyone with everything that they deem important. The contrasting precedent of the end-stage renal disease program discussed earlier may turn out to be an historical anomaly. After all, that program sprung from a fortuitous combination of technological breakthroughs, media attention, clever lobbying in Congress, and the unique cultural context of the sixties.[40] It is unlikely that similar circumstances would coincide in the case of future genetic technologies.

Moreover, there simply is no precedent for providing people with widespread access to genetic enhancements. The closest nonmedical analogy might be public education, where there seems to be a societal commitment to providing at least a basic level of education to everyone regardless of ability to pay. But even here, families that can afford the expense can send their children to superior private schools; the right to education does not entail the right to an equal education.

Perhaps, in order to refute the genetic doomsday scenario, one need not assume that everyone will have access to the genetic technologies that they desire. The division in society between the genetic have's and have not's may be far less clear-cut than has been suggested. Instead of a two-tiered society comprised of a genetic upper class and an underclass, a complex arrangement is likely to develop in which a large genetic middle class acts as a buffer and stabilizer between the upper and lower classes.

One way this might occur would be if sufficient genetic benefits were distributed to the middle class to persuade it to support a system that denied many genetic technologies to the underclass. A prime vehicle for making some genetic technologies available to the middle class but not to the underclass would be through employment benefits, in much the same way that the middle class now obtains health insurance. However, employers are likely to provide access to genetic technologies only in ways that benefit the firm. For example, they are likely to focus on providing access to technologies that increase employee productivity. Employers also would favor approaches that tended to increase employee loyalty. For example, they would be unlikely to provide younger employees with access to germ cell manipulations, which would automatically confer genetic benefits on subsequent generations, even if the employees left the company. Rational employers would prefer genetic interventions that required periodic renewal, such as the early aerosol gene therapy for cystic fibrosis that had to be repeated periodically to maintain its therapeutic efficacy. This way, the employer could hold the threat of losing the genetic benefit over the employee's head. Employees would end up being indentured to their firms in a genetic counterpart to the job-lock created by the threat of losing health insurance or coverage for preexisting conditions. It is unlikely that unions would be any more successful in regulating these employer practices than they are now at preventing the loss of health insurance if an employee leaves the company.

Even if a genetic middle class arose between the upper and lower classes, it is unclear that such a society would be able to forestall the socially destructive consequences of selective access to genetic technologies. We have assumed that the middle class would be co-opted by the genetic upper class. But it also potentially might ally with the underclass to force the upper class to disgorge more of its genetic privileges. The result could be a genetic tug-of-war that constantly threatened to degenerate into class warfare. Perpetuating such a system would take a deft hand at the controls of the mechanisms for allocating access to genetic technologies. The upper class might find it simpler to forestall an alliance between the middle and lower classes by resorting to undemocratic political means. Or the middle class might simply decide to supplant the upper class and attempt to gain genetic ascendancy for itself.

The point is that it cannot be assumed that society somehow *automatically* will figure out how to allocate access to genetic technologies in such a way that the socially disruptive scenarios we have envisioned will be avoided, much less that the net effect of the development of genetic technologies will be socially beneficial. Instead, we need to investigate whether there are any steps that might be taken to minimize the chances that the Human Genome Project will be socially disruptive. Rather than leaving the genetic future to the unplanned interaction of social forces, we need to use the political process to increase the chances that our genetic future will be democratic and harmonious. The next chapter explores the options before us.

7
Responding to the Challenge

The previous chapter described what might happen in the future if genetic technologies were distributed in the manner in which society now provides access to medical technologies. We predicted that the distribution of genetic benefits might be so imbalanced that democratic society as we know it could not be sustained. Can this possibility be avoided?

As described in Chapter Five, we currently allocate access to medical technologies through a combination of governmental and private-sector activities. If the socially destructive effects of genetic technologies can be prevented, one solution might be to reduce the role of government and give the private sector greater freedom to control the allocation of scarce genetic resources.

If governmental restrictions were eliminated completely, then *only* those who could afford to purchase genetic technologies, or who were lucky enough to receive financial assistance from family, friends, or charities, would gain access to these services. This would exacerbate, rather than alleviate, the societal threats posed by denying genetic benefits to significant segments of the population.

Can charities be expected to make genetic technologies widely available to those who cannot afford them? If so, then government interference with the genetic marketplace may not be necessary.

There is historical precedent for expecting private philanthropy to provide access to new medical technologies. In the 1930s researchers developed artificial respirators to enable polio victims who could not breathe on their own to survive. The cost of these so-called "iron lungs" was beyond many people's means. Eddie Cantor and a group of celebrities formed an organization called the March of Dimes, which called upon everyone to contribute dimes to defray the costs of iron lungs and polio research. Evidently the effort was successful, for there is no record of any polio victim being denied an iron lung because of inability to pay. (Eventually the need for this type of philanthropy was largely eliminated by the development of the polio vaccine in the 1950s, and the March of Dimes turned its focus to raising funds on behalf of victims of birth defects.[1])

Would relying on charitable contributions solve the access problem for persons who needed genetic technologies? Several factors suggest that they would not. As noted earlier, genetic technologies likely will be expensive. It is difficult to imagine that charities could raise enough money to make a substantial number of them available to a large enough number of individuals to avoid the continued appearance of widespread rationing and access only by the well-to-do. This seems particularly dubious in view of the fact that donations to charity have leveled off in recent years.[2] As a percentage of gross domestic product or GDP, donations have remained constant at around 2 percent since 1986. In 1993 charitable contributions totalled $126.22 billion, which was more than in the previous year but less than the rate of inflation.[3] This total sum of private giving approximates the amount that would have been required by President Clinton's health reform package in order to provide basic health benefits to all Americans. If, as we expect, including genetic technologies in the basic benefits package would increase the costs of health care, at least when these technologies are first introduced, then even if all charitable giving were devoted to providing access to health care, there would not be enough to extend access to these technologies to everyone while also providing access to basic health benefits. In addition, if all charitable giving is diverted to health care, this would mean that no charitable funds would be available for other worthy causes, such as education, the arts, and religious institutions.

In terms of their high cost, moreover, genetic technologies are more likely to resemble organ transplants than iron lungs. The media frequently reports charitable efforts to raise money to provide organ transplants to patients who cannot afford them.[4] These efforts have differed from the iron lung experience because they focus on individual patients in need of transplants, rather than on more faceless groups of patients. Typically, the fund-raising effort has been spearheaded by the patient's family or friends. Arguably, those who were young, or attractive, or who had special connections with the media, have been more likely than others to receive the necessary funds. If organized charities stepped in to finance access to genetic technologies, they might well allocate their funds on the basis of judgments about the recipients' character, or their social worth, or on the basis of their religion or ethnicity, all of which might raise complaints of favoritism and unfairness that could lead to government intervention. As described in

Chapter Five, similar objections to the private allocation of kidney dialysis machines during the 1960s led Congress to preempt charitable access by establishing the End-Stage Renal Disease Program within Medicare.

So far, we have focused on whether genetic technologies should be distributed by the private market rather than by the government. These are not necessarily exclusive approaches. The government's role in allocating health care would be reduced, rather than eliminated. There could be fewer government subsidies to enable low-income individuals to purchase health care, such as through the Medicaid or Medicare programs. States could be given more freedom to ration access to specific technologies, such as Oregon does under its Medicaid rationing program. State laws mandating that private insurance companies provide coverage of certain specified medical services could be weakened or repealed, and exemptions from these laws, such as the one currently provided under ERISA for employer self-insured health plans, could be broadened. Finally, statutory and perhaps even constitutional limitations on restricting access, such as the Americans with Disabilities Act, the civil rights laws and the equal protection clause of the fourteenth amendment, could be removed.

While this approach might be attractive in terms of reducing government spending and bureaucracy, it invariably would increase the likelihood that genetic technologies would be available only to those persons who could pay for them. This would increase, rather than decrease, the social problems that were identified in the previous chapter.

The key to reducing the risk of social disruption from genetic technologies, then, lies not in relaxing governmental controls on access to genetic technologies, but in relying on the appropriate use of governmental power. The question is, what form should governmental intervention take?

BANNING GENETIC TECHNOLOGIES

One alternative would be to ban access to genetic technologies for everyone. If the risks to society of genetic stratification were perceived to be great enough, and if the high cost of genetic technologies precluded the government from subsidizing access for those who were not wealthy enough to afford them, then one solution would be for the government to prevent *anyone* from obtaining access to the genetic technologies.

As a general approach to allocating scarce resources, this notion finds support among some ethicists. Edmond Cahn's apparent solution to the "lifeboat problem," in which a lifeboat is so overloaded with people that it will capsize and all will drown unless someone is thrown out, is to let it capsize.[5] If everyone cannot be saved, in Cahn's view, then no one should be saved.

Similarly, Amy Gutmann has advocated a one-class health care system which "does not permit the purchase of health care to which other similarly needy people do not have effective access."[6] Gutmann belongs to a group of philosophers called *egalitarians*. Egalitarian theories hold that every person is to be treated equally because we are all fundamentally equal.[7] To say that we are fundamentally equal is to say that everyone in a society is entitled to equal respect and equality of opportunity.[8] Applying this idea to the provision of health care services means that every person who has the same type and degree of health care need must be given an appropriate treatment if any person receives that treatment.[9] This is known as the *equal access principle*. Thus, society may provide a health care service if, and only if, it makes the service available to everyone who needs it. According to the equal access principle, it is unjust to treat some if we cannot treat all.

The values underlying the equal access principle as applied to health care include equal respect, equal opportunity, and a claim to equal treatment when physical pains reach a sufficient level.[10] That is, everyone has an equal right to be free from pain, suffering, disability, and premature death because each has intrinsic worth; health care services directed at alleviating these pains must be equally available to all who suffer them.[11] In order to ensure that everyone either received the same treatment or had an equal chance of receiving it, those who had the personal resources to purchase the treatment would be forbidden to do so.

This approach has a medical precedent. Once again, it comes from the dialysis crisis of the 1960s. At least one hospital, finding itself unable to provide free treatments to all in need, discontinued its dialysis program rather than treat only those who could pay.[12]

An interesting question is how such a ban would be effectuated. One option would be for the government to halt funding for the

Human Genome Project. Without the necessary research funds, scientists would be unable to continue to unlock the secrets of the human genome. If the location and sequence of the genes responsible for the traits that would characterize and sustain a genetic upper class remained undiscovered, it would be impossible to employ genetic technology to create and perpetuate its membership.

There is precedent for this, too. The government only recently lifted a prohibition against federal funding of medical research using tissue from aborted fetuses, which had been in effect since 1988.[13] A similar prohibition persists on research using human embryos, although a panel of experts has urged that the ban be lifted in the case of certain research using unwanted embryos that are left over from *in vitro* fertilization efforts.[14]

In fact, a research prohibition might not even have to be established by the government. In 1975 a group of scientists and other experts convened a conference at Asilomar, California, to consider whether experiments on recombinant DNA should be undertaken in light of the risks of unintentionally releasing lethal organisms, which were believed to be largely unknown and potentially devastating, into the environment. The conference concluded by calling for a voluntary moratorium on certain recombinant DNA experiments that might release dangerous genetically altered organisms into the environment.[15] This moratorium lasted until 1978, when the National Institutes of Health Recombinant DNA Advisory Committee issued new guidelines to review and approve federally funded research in the field.[16] Still, for some time after that, the government refused to approve studies involving gene therapy in humans.

While a ban on federally funded genetic research would be fairly easy to establish from an administrative standpoint, it is difficult to imagine government policymakers being willing to pay the price of forgoing the benefits from future genetic technologies. After all, a research ban would affect the mapping and sequencing of genes associated with dread diseases and inherited disabilities. What administration would be able to withstand pressures from physicians and patient groups to continue funding research to identify the genes for cancers, heart disease, mental illness, and the like, particularly if, as we predict, the mapping and sequencing of these genes may pave the way for genetic interventions that prevent or ameliorate the symptoms of these disorders?

Even if the government were to halt the Human Genome Project, the commercial potential of its technologies makes it likely that private corporations would continue to carry on the research, albeit at a slower pace. The only alternative would be for the government to make privately funded genetic research illegal.

The federal government does prohibit some types of private biomedical research. Research on human subjects that is conducted without proper safeguards can lead the research institution and the investigators to be disqualified from receiving federal research funds in the future.[17] The Food and Drug Administration prohibits companies that sponsor human trials of drugs and medical devices without obtaining prior government permission from marketing the products.[18]

Yet the commercial value of genetic technologies is liable to spawn an illicit research effort in defiance of government controls. Considerable unregulated research takes place on illegal drugs, for example. Whether the government could interdict research on genetic technologies would depend in large part on the type of research and the resources needed to carry it out. Small-scale studies on a few individuals, using commonly available equipment and materials, might be difficult to detect. Large-scale trials or experiments that required prolonged use of exotic facilities might be easier to spot. But even this would entail a fairly extensive government surveillance program, which might be difficult to finance given budgetary limits and objections from individuals and organizations who favored genetic research. Even if the United States banned genetic research, what would prevent other countries from sponsoring or condoning it? When the U.S. government refused to sanction human gene therapy experiments, for example, an American physician carried out such an experiment in Italy. It is conceivable that the United Nations or treaties between individual governments could establish an international prohibition on genetic research, but it is hard to imagine that such an arrangement could be policed effectively. In any event, in view of international economic competition and the potential enormous benefits from genetic technologies, the United States is unlikely to let foreign countries obtain a decisive advantage in such a lucrative and important field. Imagine the uproar that would erupt over a "gene drain" or "gene gap" if American scientists fled overseas or if other countries became leaders in genetic technologies.

As an alternative to a total ban on genetic research, perhaps the government merely could prohibit research on certain technologies, such as germ cell engineering or genetic enhancements, on the grounds that they created the gravest threat to democratic institutions. Aside from the problems already mentioned that would arise with any ban on scientific research, it may be difficult to isolate enhancement research from research in permissible areas, such as gene therapy for disease. The techniques for mapping, sequencing, and manipulating disease genes and genes for other traits are similar. Decoding the entire human genome will reveal the location and makeup of all genes, not just those for diseases and other conditions that we decided were suitable for further research. Even if a ban could target research on enhancements only, the prospect of gaining information about and ultimately manipulating nondisease genes is likely to be too tempting socially and commercially to forestall further research completely or for very long.

If research on genetic technologies in general, or on genetic enhancements in particular, cannot be blocked, then perhaps the government could prohibit their use. Physicians and others could be punished for performing genetic tests or therapy. Or, if the greater threat to society were thought to stem from genetic enhancement, the government could ban the provision of enhancement-type services, such as testing for nondisease or "normal" traits, or altering these traits. Violating these bans could lead to suspension or loss of a doctor's license to practice medicine, fines, or imprisonment. Institutions such as hospitals and health plans that provided proscribed genetic services could lose their licenses, their accreditation, or their ability to be reimbursed for services provided to enrollees in government programs such as Medicare and Medicaid. The institutions or their officers also could face fines and imprisonment. All of these penalties currently are imposed for various violations of federal and state laws governing health care providers, and at one time, the state of California imposed similar penalties for providing a test for birth defects—the test for maternal alpha fetoprotein.

If this prohibitory approach seems far-fetched, consider a recent statement by the Council on Ethical and Judicial Opinions of the American Medical Association. In one section of a set of recommendations for dealing with the ethical, legal, and social problems of prenatal genetic testing, the AMA states that genetic enhancement of fetuses and embryos should be permitted only under three conditions. Two of these are noncontroversial. First, there should be a "clear and meaningful benefit" to the fetus or child. Second, the benefit should occur without a concomitant degrading of other inherited characteristics or traits. But the third criterion is that, "there would have to be equal access to genetic technologies, irrespective of income or other socioeconomic characteristics." The AMA bases its position on the recognition that, "[i]f access depended on wealth, social divisions would widen, and the promise of equal opportunity for all citizens would quickly become an illusion."[19]

The AMA does not proceed to explain how it would accomplish its egalitarian objective. One option, as we have discussed, would be for society to provide sufficient funding to give all Americans access to genetic enhancements. But if this would be too expensive, as we suspect, then the AMA's policy would appear to require a ban on the private purchase of these technologies.

A less drastic policy might be a partial or temporary prohibition—one that restricted the private purchase of genetic enhancements until everyone had obtained access to the technologies necessary to make them "normal." Under this approach, gene therapy for genetic disorders might be permitted, but genetic enhancement would be allocated first to those individuals who were below accepted norms for various nondisease traits, such as intelligence or beauty. Once everyone were brought up to an equal level of these endowments, the government would allow private purchases of additional benefits. This would reduce the genetic distinctions between those who could afford to purchase genetic enhancements and those who could not, thereby perhaps decreasing the social concerns described in Chapter Five.

Yet such a policy still would require a massive regulatory program. The government would have to distinguish between genetic interventions that dealt with diseases and disorders and interventions that affected traits that were off-limits. It would have to determine what constituted a "normal" genetic endowment. The government then would have to test individuals to ascertain the nature and degree of the genetic improvements to which they were entitled, and arrange and pay for the services to be provided. A lingering issue would be whether individuals would receive germ cell engineering, so that their offspring would be more likely to begin life with a "normal" genetic makeup, or whether only somatic cell engineering would be provided, so that further testing and intervention might be necessary for each successive generation.

Any restrictions on the purchase of genetic technologies, including a ban on purchases by the well-off while everyone was being brought up to genetic speed, so to speak, also would risk the creation of a black market.[20] Such a market emerged for drugs to stave off or treat AIDS that had not been approved by the Food and Drug Administration, with "buyers' clubs" being formed through which illegal products were bought and sold.[21] Another example of a black market in medical procedures was the practice of abortion before the Supreme Court's 1972 decision in *Roe v. Wade.* Prior to that time, abortion was a felony in nearly every state, with penalties ranging from a one thousand-dollar fine to fifteen years imprisonment. Despite this, it is estimated that in the early '60s, 1 in every 5 pregnancies in the United States was terminated by abortion. In 1962 alone, more than one million abortions were believed to have been performed, one-half of which were by physicians.[22]

The two most drastic health-related attempts to eliminate black markets have been the prohibition against alcoholic beverages during the 1920s and 1930s, and the current War on Drugs. Neither presents an encouraging picture of success. Both efforts correlate with a dramatic increase in criminal activity. Both triggered widespread disobedience by large numbers of otherwise law-abiding citizens.

While Prohibition took place at a time when surveillance and interdiction techniques were primitive compared with those available to modern police forces, the War on Drugs has been a failure despite the use of the most sophisticated techniques available. The program was officially begun by President Reagan in 1982. Strict antidrug legislation was passed in 1984, 1986, and 1988 to supplement the basic statute, the Controlled Substances Act of 1970.[23] The campaign has been waged by the federal Drug Enforcement Agency, the Customs Service, the Coast Guard, the FBI, the military, the Civil Air Patrol, and state and local law enforcement officials.[24] Between 1981 and 1989 alone, the federal government spent approximately twenty billion dollars in the attempt to interdict illegal drugs. Nevertheless, drug use in the United States appears to have increased significantly, and the antidrug campaign has been charged with causing or contributing to a number of other social problems, including an increase in violent crime, prison overcrowding, violation of civil liberties, the destruction of the inner city, and the production of more potent and dangerous drugs.

Stifling a black market in genetic technologies could be just as difficult and expensive. Just how difficult or expensive would depend in large part on how difficult and expensive it would be to mount a successful law enforcement program. This in turn would depend on what form the genetic technologies took and how they were provided. If the technologies consisted of small, easily concealed capsules or devices, for example, black market traffic would be virtually impossible to interdict. Germ cell manipulation, on the other hand, would require *in vitro* fertilization or other reproductive techniques in a more visible clinic or hospital setting. Yet the availability of unlawful abortions prior to *Roe v. Wade*, despite the need for a fair amount of equipment, suggests that even these procedures could flourish in many places.

The experience with illegal drugs and abortions also indicates that no matter how extensively the U.S. government attempted to combat the domestic black market, it would have little control over access to genetic services in foreign countries. Individuals, therefore, might travel to foreign countries to obtain technologies that were banned in the United States, just as they obtained unlawful abortions, cancer treatments, and AIDS drugs in Mexico, the Bahamas, and Europe.

The U.S. government might pressure foreign nations to adopt and enforce the same restrictions on genetic technologies as those of the United States, but this is likely to be no more successful than U.S. efforts to prevent the manufacture and export of illegal drugs from other countries. If genetic technologies required relatively elaborate procedures, individuals might have to travel to foreign countries to obtain them. The government then might try to prevent Americans from traveling abroad to obtain genetic services. We have seen such travel restrictions before, in the form of prohibitions against travel to Cuba, Albania, China, and certain other countries.[25] Americans who violated these restrictions could lose their passports,[26] and those who attempted to travel without a passport could be fined or imprisoned.[27] The government also could punish individuals returning from abroad bearing illegal genetic materials or having obtained prohibited genetic services. Perhaps it will be possible to test people at customs check-points for the presence of genetic alterations. The courts have permitted customs officials to conduct routine medical testing, including blood tests for AIDS, on aliens seeking to enter the country.[28] Federal law also permits customs officers to routinely search people entering the country, without requiring a warrant or without having probable cause that a law has been violated.[29] However, the customs officer must have a reasonable suspicion of an unlawful act in order to conduct a strip or body-cavity search.[30] Similarly, the courts may require evidence of illegality before they would permit genetic "searches," which probably would include DNA testing. But what would happen in the case of individuals who were caught returning from abroad having obtained illegal genetic services? They might be fined or imprisoned, but the damage from genetic manipulation would still have been done. Would we attempt to reverse the genetic alterations they had received, or to sterilize them if they had obtained germ cell engineering? Would these penalties survive a constitutional challenge as cruel and unusual punishments?

A better enforcement approach might be to deny individuals the necessary funds to pay for genetic technologies abroad. Federal law currently permits the president to impose financial restrictions in peacetime on individuals and organizations during a crisis that poses an "unusual or extraordinary threat" to national security, foreign policy, or the economy.[31] These sanctions have been levied on financial transactions with Libya, South Africa, Iran, and Nicaragua. The law also authorizes the secretary of the treasury to require reports of all foreign currency transactions by U.S. citizens in excess of ten thousand dollars,[32] and to stop and search persons attempting to leave the country without filing the required reports.[33] These laws might be invoked to prevent people from transferring money abroad in order to purchase banned genetic technologies. All of these provisions have been upheld by the courts against challenges that they unconstitutionally infringed on individual rights.[34]

Nevertheless, experience with drug money laundering, Swiss bank accounts, and offshore investment companies suggests that the blockade that these legal measures would hope to achieve would be, in reality, easily breached. Financial markets increasingly are becoming global; Americans invest money abroad in a myriad of ways. It does not seem realistic to expect the government to monitor financial markets carefully enough to prevent foreign genetic purchases, except perhaps in the case of technologies that were so extraordinarily expensive that they required large transfers of funds.

But there is a more fundamental problem with all of these approaches. They assume that we would be willing to give up the enormous potential benefits that banned genetic technologies could provide. The same characteristics that would make these technologies so desirable would militate against banning them altogether. In Chapter Five, we concentrated on the threats that these technologies would pose to society, on the assumption that the benefits they yielded would accrue primarily to the wealthy individuals. But directly or indirectly, these individuals might be capable of providing substantial social benefits. Their physical capabilities could extend our reach to realms that were previously off-limits or that could only be reached remotely, like the deep sea or outer space. Their mental abilities could stimulate new arts and industries.

Furthermore, a successful ban on genetic technologies would produce the peculiar result that people could not get an essential good (medically necessary treatment), but could use their wealth to satisfy their nonessential desires (e.g., for luxury cars or vacation homes).[35]

We have a long history in this country of allowing people to purchase anything they wish, so long as doing so does not directly harm someone else. We have tolerated few exceptions to this libertarian principle. Almost without exception, prohibitions against liquor, drugs, pornography, and abortion are defended on the grounds that these practices harm third parties. In the case of the genetic technologies that we described in Chapter Five, it is hard to point to any direct harm to third parties. The threats they pose to society, while potentially extremely destructive, are likely to be too subtle and diffuse to sustain the type of crusade that would lead to a complete ban, including one that extended to foreign markets.

GENETIC HANDICAPPING

As a result of the problems created by prohibiting access to genetic technologies, we might be better off recognizing that some people will obtain access to genetic technologies, if necessary, illegally. Instead of trying to prohibit this activity altogether, we might try to mitigate its adverse effects. If, as suggested in Chapter Five, a major problem is that if access to genetic technologies were provided along the same lines as access to currently available medical services, and only the wealthy were likely to be able to afford certain powerful gene therapies and genetic enhancements, then one approach might be to try to cancel out the social advantages that these technologies provided. For example, we might place the genetically enhanced at a disadvantage in competing for certain benefits, such as jobs, public office, and contracts for goods and services. This could be achieved either by handicapping those who were enhanced, or by giving preferences to the nonenhanced. Minority set-asides and affirmative action programs currently give persons competitive advantages based on race or ethnicity. Similar programs might be instituted based on genetic ability.

This alternative has two main shortcomings. First, we would need some method of ascertaining who had obtained genetic technologies and who had not. As suggested earlier, to intercept persons returning from abroad with genetic contraband, techniques might be developed for "tagging" genetic manipulations so that they might be detected with some sort of genetic test. But more likely, we would have to rely on recordkeeping to monitor who had received access to genetic technologies. This again conjures up the possibility of a black market in which people could obtain genetic technologies anonymously or without being able to be traced.

The second problem with genetic handicapping is that it would sacrifice the social benefits from genetic enhancement. If a person who was better able to perform a job because of genetic manipulation were not hired in favor of someone who was less qualified, the job would not be done as well. This marks a critical difference between genetic handicapping and current affirmative action programs. Affirmative action is based on the assumption that there are no relevant performance differences between the person who is given a preference, say, in being hired for a job, and the person who is rejected. The notion behind affirmative action is that it cancels out the *dis*advantage that the person who is given the preference otherwise would suffer on account of, say, her race.

Genetic handicapping, on the other hand, would require us to ignore actual performance differences between individuals. We seem willing to employ handicapping in some sports, such as friendly games of golf, but how likely is it that we would accept this when something more important was at stake? Would we be willing to entrust an airliner, for example, to a pilot who had been hired over someone with better eyesight, or stamina, or quicker reflexes, simply in order to level the social playing field?

The closest we now come to a situation in which we *are* willing to overlook true, relevant, performance differences between individuals is the legal prohibition of discrimination against persons with disabilities. If a person with a disability is not able to perform a job as well as a person who is not disabled, an employer cannot necessarily refuse to hire the person with the disability. Instead, the employer must make a reasonable effort to help that person compensate for the disability, for example, by installing special equipment, even though this means that the employer must spend money that would not have to be spent in the case of a nondisabled employee. Only if, after receiving a reasonable accommodation, the employee still could not perform the job adequately, the employer could decline to hire or retain the person with the disability. In effect, then, the law requires the employer to pay more for the labor of the individual with the disability, or, to put it another way, to accept less performance for the same labor costs as for a nondisabled employee. Assuming that society was committed to denying advantages to individuals who were able to obtain access to genetic technologies, and that it was willing to incur the costs, it might pass similar laws requiring special treatment for the genetically disadvantaged.

Another method for limiting social unrest over genetic stratification might be to permit a sufficient number of individuals to obtain access to genetic enhancements irrespective of their ability to pay. One way to do this would be for the government to identify certain groups within society as "genetically disadvantaged" and to subsidize genetic technologies for the members of these groups. But allowing the government to play this role would be extremely controversial, not to mention dangerous. How would "genetically disadvantaged" be defined?

A policy of excluding coverage of genetic enhancements would not be easy to enforce. For one thing, it would be difficult to determine whether a genetic intervention was utilized merely to correct a patient's problem, or instead, to enhance the patient. Guidelines would have to be established for when a trait was within a "normal" range, and when it lay beyond. As we mentioned in Chapter Two, one way we have done this in the past is simply to calculate the statistical variation in a trait within the population, and then to decide how much variation from the mean will be deemed "normal."[36] The problem with this approach is that if people gained access to genetic enhancement despite lack of insurance coverage, the values for the characteristics that were enhanced would migrate upwards. What was once normal would begin to fall outside normal limits, and what was once excluded as enhancement would be eligible for coverage as a "correction." This would place continuing pressure on policymakers to cover genetic enhancements in borderline cases.

One solution might be to identify persons who possessed genes that were associated with below-normal characteristics or abilities, say, one or two standard deviations below the mean. But what is "normal"? Since the determination of normality is based on a statistical comparison of individuals in society, creating the infamous "bell curve," some people always will be below normal, no matter how extensively genetic intervention improves societal traits. Moreover, how should we deal with the fact that one person's notion of a disadvantage might be another person's notion of distinctiveness? Imagine the uproar if the government sought to declare race a disadvantaged genetic condition and offered people the option of altering the racial characteristics of their offspring. Concentrating resources on rectifying genetic disadvantages could distract us from confronting other sources of inequality, such as wealth or prejudice. Finally, relying on the government to designate genetic disadvantages could lead to more overt eugenics efforts, such as encouraging or forcing people to "correct" their genetic inheritances.

GENETIC LOTTERIES

Before proceeding, let us review the shortcomings of the policy options that we have discussed. We began by assuming that genetic technologies, particularly genetic enhancements and more expensive versions of gene therapy, would be scarce resources. In other words, there would not be enough of them, or enough of them at a low enough price, to provide them to everyone who desired them. As a result, under our current allocation policies, these technologies would be available primarily to those who could purchase them with personal assets, either directly or by buying high-cost insurance plans that covered these services. We rejected a pure market approach that would eliminate what few subsidies would be provided under the current allocation system to enable less well-off people to obtain access to genetic technologies; a pure market approach would only worsen the social dislocations that would result from distributing access to genetic technologies according to current access policies. We also described how unrealistic it would be to try to bring the Human Genome Project to a halt, and how difficult it would be to try to prevent the development or use of genetic enhancements and other genetic technologies that posed the greatest societal threats. Finally, we considered various ways to level the societal playing field, all of which were fraught with philosophical and practical problems.

In our opinion, the best approach would be to operate a genetic lottery. The lottery would be open to anyone, but everyone would have the same chance of winning; that is, no one could obtain a greater number of "tickets" than anyone else. The lottery would be voluntary; people who objected to genetic technology on religious or other grounds could decline to participate.

Winners would receive access to a complete package of genetic services, from which they could choose what they desired. The benefits available through the lottery would be the same as those available in the market. If wealthy people were able to purchase germ cell genetic enhancements, for example, so would lottery winners.

Of all the policy options we have discussed, a genetic lottery would go the farthest toward solving the societal problems that we have identified. It would accommodate the assumption that not all genetic services could be made available to everyone. It would permit continued research toward conquering genetic diseases. It would enable people, who could not otherwise afford them, to obtain access to genetic technologies. Within the lottery, at least, there would be true equality of opportunity. By permitting winners to obtain access to whatever genetic services they wished, so long as the services were available on the open market, the lottery would avoid the onerous task of deciding the technologies to which people should be given access. A lottery would reduce the demand for black market genetic technologies, since there would be less need to restrict the availability of these technologies on the open market in order to promote fairness. Instead, the government could focus its regulatory efforts on ensuring that the technologies that were available were safe and effective, and that they were delivered by adequately skilled health care professionals.

A genetic lottery would have a number of other advantages. The odds could be adjusted to make winning easier or more difficult. If the costs of genetic technologies increased, the odds of winning could be reduced to save money. If polls or other public opinion indicators showed that there was a significant increase in the number of people opposed to wealth-based access to genetic benefits, the odds could be "sweetened" so that enough people won to abate the discontent.

Another advantage of a lottery is that not only would the winners be selected at random, but so would their genes. A wider pool of genes would flourish because their hosts would receive gene therapies and genetic enhancements. This would promote genetic diversity and enrich the gene pool available for future generations.

As everyone knows, state lotteries are widely used in the United States to raise money for education and other public purposes. But they serve another important function in that they create a possibility of sudden, dramatic upward mobility, and by providing a random chance of winning, they help preserve the notion of equal opportunity that is fundamental to Western democratic society. Poor people view the lottery as "the only possibility for breaking out of the cycle of poverty they live in."[37] Thus, lotteries can be an effective means of reducing lower class discontent with the growing gap between rich and poor, or, in our case, the gap between the genetic aristocracy and the rest of society.

Current state lotteries raise a number of objections. Studies show that tickets for state lotteries are purchased primarily by the poor. In Maryland, for example, half of the tickets are purchased by the poorest third of the population. Nationally, one-third of families with less than ten thousand dollars in annual income spend 20 percent of their income on lottery tickets. As a mechanism for financing state and local governments, lotteries are highly regressive. Moreover, the odds of winning are so low that the prospect of upward mobility, as one commentator put it, is really only "the sale of an illusion to poor people,"[38] who can ill afford the price.

The type of genetic lottery we have in mind would avoid these problems. Chances would not be for sale. The government would administer the lottery by giving each adult a chance, unless the adult chose not to participate. Poor people could not waste what little money they had by buying tickets. The lottery would not be used as a means of raising public funds.

Lotteries have had a long history for a variety of purposes. As Barbara Goodwin recounts in her book, *Justice By Lottery*,[39] lotteries were used by the ancient Greeks and, to a lesser extent, the Romans, to allocate political duties. The Roman emperors gave away prizes by lot during festivals. Queen Elizabeth I established a lottery in 1566 to raise money to build a harbor, and the English government continued to use it to obtain revenue until the early nineteenth century, and reinstituted it in 1992. New Hampshire introduced the first state lottery in the United States in 1964. Federal courts in the United States have approved choice by lot as a constitutionally acceptable method of allocating scarce public housing[40] and liquor licenses.[41]

State- and manufacturer-sponsored lotteries are already playing a role in the distribution of scarce health care resources. In 1990 the Pennsylvania state mental hospital system established a lottery program to distribute the drug Clozapine, introduced for the treatment of schizophrenia.[42] In 1993 Berlex Laboratories introduced Betaseron (interferon beta 1b), a genetically engineered form of a chemical involved in the body's immune system.[43] Because of limited supplies of recombinant interferon, Berlex utilized a random selection process to determine who would be able to begin treatment, until production capacity could meet demand, which occurred in 1995. Because of limited supplies prior to receipt of final FDA new drug approval, the same process has been used to distribute Invirase, a protease inhibitor designed to reduce the amount of the AIDS virus in the blood, and currently is being used to distribute myotrophin, the drug designed to treat patients with amyotrophic lateral sclerosis.

Choosing by lot also has been recognized as a just means of selecting members of a group to undergo unavoidable privations. The classic examples are "lifeboat" cases, where someone must be thrown overboard from a vessel at sea to enable the rest of the passengers to survive. The earliest recorded case, of course, was that of Jonah, who was cast overboard to save the ship on which he was traveling, and ended up in the whale.[44] In 1842 a U.S. court considered murder charges against a seaman who threw passengers from a sinking longboat.[45] All male passengers in the boat, except two married men and a boy, were drowned; all nine crewmen, including the defendant, survived. In the course of upholding the defendant's conviction for manslaughter on the basis that the sailors were wrong to sacrifice passengers' lives to save their own, the court said that they should have selected by lot those who were to die, stating that "[w]e can conceive of no mode so consonant both to humanity and to justice, and the occasion, we think, must be peculiar which will dispense with its exercise." [46]

A key issue would be how to finance a genetic lottery. Since it would be essential to prevent poor people from being exploited, and to give everyone an equal chance, financing the lottery by selling chances is not an option. One possibility would be to finance the lottery with general tax revenues. This would promote economic fairness, since, except for sales and other excise taxes, taxes tend to be collected on a progressive basis; those who earn or possess more generally pay more. Moreover, since genetic advantages are likely to translate into social and economic benefits, a genetic lottery funded by a progressive income tax is likely to transfer benefit from the genetically advantaged to the genetically disadvantaged.

A more straightforward way to use the lottery to level the genetic playing field would be to tax genetic technologies and use the revenues to finance the lottery. The tax could be imposed on manufacturers; distributors; providers of genetic services, such as hospitals, clinics or physicians; insurers who covered them; and even on those who received them. (The lower down the chain of distribution the tax was assessed, the larger the number of taxable events, and the more difficult it would be to collect the tax.) The amount of the tax could be adjusted depending on whether society desired to increase or decrease the odds of winning the genetic lottery. If more winners were deemed necessary in order to appease the genetic underclass, then the tax could be increased. If society felt that greater benefits could be obtained from a larger genetic upper class, and, therefore, were willing to tolerate greater genetic stratification, the tax could be reduced.

The idea of taxing health benefits in order to subsidize access to health care for the less well-off is not new. It was a major feature of many of the health reform proposals that were introduced in Congress in 1993–1994, including that of the Clinton administration. [47] A tax on genetic technologies also would tend to restrain the purchase of these technologies by those who could afford them. By reducing the size of the genetic upper class, or the genetic gap between it and the genetic underclass, the tax might reduce the friction between genetic have's and have not's.

8
Conclusion

Earlier in this book, we distinguished between the "bads" that might flow from the Human Genome Project—the threats to happiness and well-being from genetic testing, discrimination, or eugenics—and the "goods"—the benefits to individuals and society from ameliorating or preventing genetic disorders and from improving upon natural genetic endowments. We said that, contrary to most of the social commentary being written about the Human Genome Project, we were going to concentrate on the issues raised by the "goods"—how individuals can obtain the benefits of the genetic revolution.

In the end, however, we find ourselves heralding another set of worries—that individuals will not be able to obtain access to the new genetic technologies without being able to pay for them and that the resulting maldistribution of genetic advantages will threaten the fundamental principles upon which Western democratic societies are based.

Some will say that the genetic advances that give rise to our concerns are so unlikely or so far in the future that society should not waste time on this subject, and that focusing on it only diverts attention from far more pressing social problems. Several years ago, for example, we attempted to publish an article on allocating access to genetic technologies in a prestigious medical journal, and we were told that the article would be accepted only if we deleted all references to genetic engineering and enhancement.

This view is dangerously naive. Scientists have many exquisitely difficult challenges to overcome before some of these genetic technologies become realities. But as we pointed out in Chapter Two, presently there is no reason to believe that researchers will encounter insurmountable technical hurdles that will block further progress. Given the state of science, there is a good chance that most of the genetic technologies that give rise to our concerns will be developed within our lifetimes.

As we acknowledged in Chapter Six, it is possible that, contrary to our expectations, society will adapt well enough to genetic technologies that unequal access will not be a serious problem. Alternatively, ethical and religious opposition to genetic engineering and enhancement technologies may be so great that no one will seek to avail themselves of the benefits.

Any of these might occur. But are we willing to take the risk of being unprepared in case they do not? Society may have more immediate problems, but we cannot identify any that are more serious.

We suspect that one reason that we were criticized for discussing access to genetic engineering or enhancement when we tried to publish our article is that the genetic research community is afraid of what might happen if nonscientists began to take our concerns seriously. The public might clamor for the Human Genome Project to be shut down, or at least for research that might lead these developments to be stifled. Genetic researchers may fear that they would be prevented from pursuing less controversial objectives, such as developing methods of identifying and treating genetic disorders.

We acknowledge this risk, but we feel that it is small and worth taking. The Human Genome Project is not about to be abandoned so long as it holds out the prospect of conquering genetic disease. We are not in any way advocating that public funding for this research be cut off or reduced.

At the same time, society must make certain that these technologies become available in ways that do not destroy the democratic system. We have suggested one way that this might be prevented. There may be better solutions that may be advanced by others. But it is clear that ignoring these issues now, and dealing with the access question on a laissez faire basis when and if it arises, is foolhardy. We are raftsmen fast approaching a social and evolutionary maelstrom. Whether we will emerge safely will depend on how well prepared we are, not to mention a great deal of luck.

References

CHAPTER 2: THE HUMAN GENOME PROJECT

1. Lewis Thomas, "The Technology of Medicine," *The Lives of a Cell: Notes of a Biology Watcher* (New York: Viking, 1974), 35–42.
2. Robert M. Blaese, et al., "Treatment Of Severe Combined Immunodeficiency Disease (SCID) Due To Adenosine Deaminase Deficiency with CD34+ Selected Autologous Peripheral Blood Cells Transduced with a Human ADA Gene," *Human Gene Therapy* 4 (1993): 521–27.
3. Edward H. Oldfield, et al., "Gene Therapy for the Treatment of Brain Tumors Using Intratumoral Transduction with the Thymidine Kinase Gene and Intravenous Ganciclovir," *Human Gene Therapy* 4 (1993): 39–69.
4. Philip Leder, David A. Clayton, and E. Rubenstein, eds., *Introduction to Molecular Medicine*, (New York: Scientific American Books, Inc., 1994); James D. Watson, John Tooze, and David T. Kurtz, *Recombinant DNA: A Short Course*, (New York: Scientific American Books, Inc., 1983); D.J. Weatherall, *The New Genetics and Clinical Practice*, (New York: Oxford University Press, 1991).
5. Institute of Medicine, Committee on Assessing Genetic Risks, *Assessing Genetic Risks: Implications for Health and Social Policy* (Washington, D.C.: National Academy Press, 1994), 21–22.
6. United States Congress, Office of Technology Assessment, *Mapping Our Genes–The Genome Project: How Big, How Fast?* OTA-BA-373 (Washington, D.C.: U.S. Government Printing Office, 1988).
7. R. Lipkin, "The Quest to Break Human Genetic Code," *Insight* (Dec/Jan 1991): 46–48.
8. James F. Gusella, et al., "A Polymorphic DNA Marker Genetically Linked to Huntington's Disease," *Nature* 306 (1983): 234.
9. The Huntington's Disease Collaborative Research Group, "Novel Gene Containing a Trinucleotide Repeat That Is Expanded and Unstable on Huntington's Disease Chromosomes," *Cell* 72 (1993): 971.
10. Eliot Marshall, "A Strategy for Sequencing the Genome 5 Years Early," *Science* 267 (1995): 783–784.

CHAPTER 3: THE PRACTICAL APPLICATIONS OF HUMAN GENETIC TECHNOLOGY

1. Jean L. Marx, "Gene Signals Relapse of Breasts, Ovarian Cancers," *Science* 244 (1989): 654–55.
2. Karin L. McGowan, "Infectious Diseases: Diagnosis Utilizing DNA Probes," *Clinical Pediatrics* 28 (1989): 157–62.

3. Yoshio Miki, et al., "A Strong Candidate for the Breast and Ovarian Cancer Susceptibility Gene BRCA1," *Science* 226 (1994): 66–71.

4. David Sidransky, et al., "Identification of *Ras* Oncogene Mutations in the Stool of Patients with Curable Colorectal Tumors," *Science* 256 (1992): 102–5.

5. Alan H. Handyside, et al., "Birth of a Normal Girl after In Vitro Fertilization and Preimplantation Diagnostic Testing for Cystic Fibrosis," *New England Journal of Medicine* 327 (1992): 905–9.

6. Larry Thompson, "At Age 2, Gene Therapy Enters a Growth Phase," *Science* 258 (1992): 744–46.

7. Barbara J. Culliton, "Endothelial Cells to the Rescue," *Science* 246 (1989): 749.

8. Stephen A. Rosenberg, "Gene Therapy for Cancer," *Journal of the American Medical Association* 268 (1992): 2416–19.

9. Takeshi Ohno, et al., "Gene Therapy for Vascular Smooth Muscle Cell Proliferation after Arterial Injury," *Science* 265 (1994): 781–84.

10. Mario R. Capecchi, "Altering the Genome by Homologous Recombination," *Science* 244 (1989): 1288–92.

11. See Nelson A. Wivel and LeRoy Walters, "Germ-Line Gene Modification and Disease Prevention: Some Medical and Ethical Perspectives," *Science* 262 (1993): 533–38; Ronald Muson and Lawrence H. Davis, "Germ-Line Gene Therapy and the Medical Imperative," *Kennedy Institute of Ethics Journal* 2 (1992): 137–58; Colin Norman, "Clerics Urge Ban on Altering Germ Cells," *Science* 220 (1983): 1360–61.

12. Klaus Lindpaintner, "Finding an Obesity Gene: A Tale of Mice and Men," *New England Journal of Medicine* 332 (1995): 679–80.

13. Yiying Zhang, et al., "Positional Cloning of the Mouse Obese Gene and Its Human Homologue," *Nature* 372 (1994): 425–32.

14. Jeffrey L. Halaas, et al., "Weight-Reducing Effects of the Plasma Protein Encoded by the Obese Gene," *Science* 269 (1994): 543–46.

15. Marcia Barinaga, "'Obese' Protein Slims Mice," *Science* 269 (1995): 475–76.

16. Francis Collins and David Galas, "A New Five-Year Plan for the U.S.: Human Genome Project," *Science* 261 (1993): 43; Nicholas Wade, "Rapid Gains Are Reported On Genome," *New York Times*, 28 September 1995, p. 24A.

17. Natalie Angier, "Gains Made in Effort to Map the Human Genetic Makeup," *New York Times*, 1 October 1992, p. 1A; Gina Kolata, "Scientists Speedily Locate a Gene that Causes Breast Cancer," *New York Times*, 21 December 1995, p. 18B; Gina Kolata, "Doctors Isolate a Common Cancer-Related Gene," *New York Times*, 23 June 1995, p. 1A; Rick Weiss, "Studies Link Colon Cancer to 2nd Gene," *Washington Post*, 17 March 1994, p. 1A; Associated Press, "Third Gene Tied to Early Onset Alzheimer's," *New York Times*, 19 August 1995, p. 12A.

18. James D. Watson and Norton Zender, "Genome Project Maps Paths of Diseases and Drugs," *New York Times*, 28 September 1995, p. 24A.

19. Ronald Dworkin reports, for example, that John Maddox, the editor of the prestigious journal *Nature*, scoffed at the idea of genetic enhancement because, in his opinion, it was at least twenty-five years away.

20. Rachel Bird, "Health: This is the Rest of Your Life," *The Guardian* (London), 7 January 1997, p. 16T; Gina Kolata, "Breaking Ranks, Lab Offers Test to Assess Risk of Breast Cancer," *New York Times*, 1 April 1996, p. 1A; Gina Kolata, "Tests to Assess Risks for Cancer Raising Questions," *New York Times*, 27 March 1995, p. 1A; Richard Saltus, "Cystic Fibrosis Test Stirs Wonder, Worry," *Boston Globe*, 21 October 1990, p. 2.

21. Thomas Stuttaford, "Hormone Can Add to Your Height," *Times* (London), 12 September 1996; Marjorie Sun, "Gene-Spliced Hormone for Growth Approved," *Science* 230 (1985): 523; Gina Kolata, "New Growth Industry in Human Growth Hormones," *Science* 234 (1986): 22.

22. Stuart Auerbach, "Is Human Growth Hormone Overprescribed?" *Washington Post*, 27 August 1996, p. 7Z; Stuart Auerbach, "Prescribing Hormones in Pursuit of Taller Kids," *Bergen Record*, 9 September 1996, p. 1H; Sandra Blakeslee, "Supply of Growth Hormone Brings Hope for New Uses," *New York Times*, 10 February 1986, p. 1C.

CHAPTER 4: THE IMPACT OF GENETIC TECHNOLOGIES

1. See, for example, Jeremy Rifkin and Nicanor Perlas, *Algeny: A New Word, A New World* (New York: Penguin Books, 1984).

2. Judith A. Boss, *The Birth Lottery: Prenatal Diagnosis and Selective Abortion* 22 (Chicago: Loyola University Press, 1993).

3. Ibid., 57–58; 155.

4. Gene Levinson, et al., "Recent Advances in Reproductive Genetic Technologies," *Bio/Technology* 13 (1995): 968.

5. Ibid.

6. See letter from Ronnie Blumenthal, Acting Director of Communications and Legislative Affairs, EEOC, to Rep. Bob Wise, Chairman, House Subcommittee on Government Information, Justice and Agriculture, 22 November 1991, responding to questions submitted to the EEOC.

7. *EEOC Compliance Manual* (CCH) § 905 (March 15, 1995).

8. See Americans with Disabilities Act, Title V, 42 U.S. Code § 12201 [§501](c).

9. Ibid.

10. *Americans with Disabilities Act Regulations and Interpretive Guidance: EEOC Interim Guidance on Application of ADA to Health Insurance*, 2 Empl. Prac. Guide (CCH) § 5375 (June 8, 1993).

11. National Institutes of Health Task Force on Genetic Information and Insurance, *Genetic Information and Health Insurance: Report of the Task Force*, NIH-DOE Working Group on Ethical, Legal and Social Implications of Human Genome Research, 1993.

12. See, for example, California Insurance Code § 10148 (West 1995); Colorado Revised Statutes § 10-3-1104.7 (1995); Georgia Code Annotated § 33-54-1 (1996); Minnesota Statutes § 72A.139 (1995); New Hampshire Revised Statutes Annotated § 141-H:4 (1995); Ohio Revised Code Annotated §§ 1742.42 & 1742.43 (Baldwin 1996); Oregon Revised Statute § 746.135 (1995); Wisconsin Statutes § 631.89 (1994).

13. Roger Doughty, *The Confidentiality of HIV-Related Information: Responding to the Resurgence of Aggressive Public Health Interventions in the AIDS Epidemic*, 82 Cal. L. Rev. 113, 132–33 (1994); William J. Kassler, et al., "Anonymous vs. Confidential HIV Testing in N.C.," International Conference on AIDS 8 (1992): C379.

14. Institute of Medicine, Committee on Assessing Genetic Risks, *Assessing Genetic Risks: Implications for Health and Social Policy* (Washington, D.C.: National Academy Press, 1994), 21–22.

15. See Daniel Callahan, *Setting Limits: Medical Goals in an Aging Society* (Washington, D.C.: Georgetown University Press, 1995); Willard Gaylin and Bruce Jennings, *The Perversion of Autonomy: The Proper Uses of Coercion and Constraints in a Liberal Society* (New York: Free Press, 1996).

16. Institute of Medicine, *Assessing Genetic Risks*, 21–22.

17. Linus Pauling, *Reflections on the New Biology: Foreword*, 15 UCLA L. Rev. 267, 268 (1986).

18. Owen D. Jones, *Sex Selection: Regulating Technology Enabling the Predetermination of a Child's Gender*, 6 Harv. J.L. & Tech. 1 (1992); Christopher Farley, "The Debate Over Uses of Prenatal Testing," *USA Today*, 2 February 1989, p. 5D.

19. Henry K. Beecher, "Ethics and Clinical Research," *New England Journal of Medicine* 74 (1966): 1354–1360; James H. Jones, *Bad Blood: The Tuskeegee Syphilis Experiment* (New York: Free Press, 1993); Advisory Committee on Human Experiments, *Human Radiation Experiments: Final Report* (Washington, D.C.: Government Printing Office, 1995); Allen Buchanan, "Judging the Past: The Case of the Human Radiation Experiments," *Hastings Center Report* 26 (1996): 25–30.

20. John Drury Ratcliff, *Yellow Magic: The Story of Penicillin* (New York: Random House, 1945); Ronald Hare, *The Birth of Penicillin and the Disarming of Microbes* (London: Allen & Unwin, 1970).

21. Stuart Auerbach, "Prescribing Hormones in Pursuit of Taller Kids," *Bergen Record*, 9 September 1996, p. 1H; Gina Kolata, "New Growth Industry in Human Growth Hormone?" *Science* 234 (1986): 22; Jeff Barr, "Long, Short of Growth Hormone," *Detroit News*, 16 November 1995, LifeM.

22. Stuart Auerbach, "Is Human Growth Hormone Overprescribed?" *Washington Post*, 27 August 1996, p. 7Z; Richard Stone, "NIH to Size Up Growth Hormone Trials; National Institutes of Health Responds to Jeremy Rifkin's Concerns," *Science* 257 (1992): 739.

23. Jane E. Brody, "Restoring Ebbing Hormones May Slow Aging," *New York Times*, 18 July 1995, p. 5B.

24. Susan O'Hara, *Comment: The Use of Genetic Testing in the Health Insurance Industry: The Creation of a 'Biologic Underclass,'* 22 Sw. U. L. Rev. 1211

(1993); T.H. Cushing, *Should There Be Genetic Testing in Insurance Risk Classification?* 60 Def. Counsel J. 249 (1993); Larry Gostin, *Genetic Discrimination: The Use of Genetically Based Diagnostic and Prognostic Tests by Employers and Insurers,* 17 Am. J.L. & Med. 109 (1991).

CHAPTER 5: ACCESS TO GENETIC TECHNOLOGIES

1. *See* Maxwell J. Mehlman, *Rationing Expensive Lifesaving Medical Treatments,* 1985 Wis. L. Rev. 239 (1985).

2. James Childress, *Triage in Neonatal Intensive Care: The Limitations of a Metaphor,* 69 Va. L. Rev. 547, 551–52 (1983); Note, *Scarce Medical Resources,* 69 Colum. L. Rev. 620, 664 n.241 (1969).

3. United Network for Organ Sharing, *Update,* January 1996.

4. Wade Ruosh, "New Ways to Avoid Organ Rejection Buoy Hopes: Xenotransplantation," *Science* 270 (1995): 234; Philip J. Hilts, "Success in Tests of Pigs' Hearts in Baboons," *New York Times,* 1 May 1995, p. 13A; Lawrence K. Altman, "Man Seems to Reject Baboon Liver," *New York Times,* 16 January 1993, p. 6A; Associated Press, "Woman Gets Liver from a Pig But Dies," *New York Times,* 13 October 1992, p. 6C.

5. Jay Katz and Alexander M. Capron, *Catastrophic Diseases: Who Decides What?* (New York: Russell Sage Foundation, 1975), 36–37.

6. Richard Rettig, *The Policy Debate on Patient Care Financing for Victims of End-Stage Renal Disease,* 40 Law & Contemp. Probs. 201, 202 (1976).

7. Telephone interview with Ira Greifer, M.D., Medical Director, National Kidney Foundation, 28 June 1983.

8. See Roger W. Evans, et al., "Implications for Health Care Policy: A Social and Demographic Profile of Hemodialysis Patients in the United States," *Journal of the American Medical Association* 245 (1981): 487–490.

9. Rettig, "The Policy Debate."

10. 21 C.F.R. § 312.34 (1995).

11. Baruch A. Brody, *Ethical Issues in Drug Testing, Approval, and Pricing: The Clot-Dissolving Drugs* (New York: Oxford University Press, 1995), 173.

12. Mary Murray, "Nancy Wexler," *New York Times,* 13 February 1994, p. 28C; "Genetics Hallmark at Sara Lawrence," *New York Times,* 3 July 1994, p. 11WC.

13. See Note, "Scarce Medical Resources."

14. James F. Blumstein, "Effective Health Planning in a Competitive Environment," in *Cost, Quality and Access in Health Care,* eds. Frank A. Sloan, James F. Blumstein, and James M. Perrin (San Francisco: Jossey-Bass, 1988), 21–43.

15. Ibid.

16. See V. Parsons and P. Lock, "Triage and the Patient with Renal Failure," *Journal of Medical Ethics* 6 (1980): 173–176.

17. Mackenzie Carpenter, "Pitt Is Proud Giant That Some Resent," *Pittsburgh Post-Gazette,* 2 February 1994, p. 1A.

18. See Note, "Scarce Medical Resources," 620.

19. Shana Alexander, "They Decide Who Lives, Who Dies," *Life* 53 (9 November 1962): 102–4.

20. Guido Calabresi and Philip Bobbitt, *Tragic Choices* (New York: Norton, 1978), 187.

21. Robbins and Robbins, "The Rest Are Simply Left to Die," *Redbook* (November 1967): 132–33.

22. David Sanders and Jesse Dukeminier, Jr., *Medical Advance and Legal Lag: Hemodialysis and Kidney Transplantation*, 15 UCLA L. Rev. 357, 378 (1968).

23. Evans, "Implications for Health Care Policy," 490.

24. Boyce Rensberger, "Pennsylvania's Gov. Casey Has Heart-Liver Transplant," *Washington Post*, 15 June 1993, p. 1A.

25. *Ibid.* Also see Ross Daly, "Transplants: National Guidelines Developed for Patients," *The Memphis Commercial Appeal*, 1 August 1993, p. 3C.

26. Claudia Coates, "Casey's Quick Transplant Renews Ethics Debate," *Los Angeles Times*, 25 July 1993, p. 3A.

27. Karen Brandon, "Furor Over Transplants for Death Row Inmates," *Chicago Tribune*, 1 March 1996, p. 12.

28. Alvin Moss and Mark Siegler, "Should Alcoholics Compete Equally for Liver Transplantation?" *Journal of the American Medical Association* 265 (1991): 1295.

29. Health Care Financing Administration, "Medicare Program: Criteria for Medicare Coverage of Adult Liver Transplants," 55 *Federal Register* 3545 (1991).

30. Letter from Louis Sullivan, Secretary of Health and Human Services, to Barbara Roberts, Governor of Oregon, 3 August 1992.

31. Proposition 187, Cal. Gov't. Code § 53069.65 (West 1995); Cal. Health & Safety Code § 130 (West 1995).

32. Celia Dugger, "Welfare States: Why Lump Sums Mean Some Lumps," *New York Times*, 28 May 1995, p. 6; Esther B. Fein, "Restructuring Medicare and Medicaid Will Affect the Low-Income Elderly Twice," *New York Times*, 5 November 1995, p. 22.

33. Katherine R. Levit, et al., "National Health Expenditures, 1995," *Health Care Financing Review* 18 (1997): 200.

34. Ibid.

35. Jane Bryant Quinn, "Health-Care Debate Gone, but Problem Isn't," *Fort-Worth Star-Telegram*, 23 August 1995, p. 2.

36. Edwin Chen, "Health Funds Good News for Kids," *Los Angeles Times*, 3 August 1997, p. A26. The Balanced Budget Act of 1997 contains a $24 billion state block grant to provide health insurance coverage over five years to up to half of the uninsured children of the working poor. Ibid.

37. See John Z. Ayanian, et al., "The Relation Between Health Insurance Coverage and Clinical Outcomes Among Women with Breast Cancer," *New England Journal of Medicine* 329 (1993): 326.

38. This situation has been addressed by the Health Insurance Portability and Accountability Act of 1996, Pub. L. No. 104–191, 110 Stat. 1936 (1996) preventing insurers from denying coverage to individuals with pre-existing conditions, although allowing insurers to impose up to twelve month waiting periods before having to pay for medical treatments associated with the pre-existing condition. This act protects employees when they change jobs, since they can neither be denied new coverage, nor can a waiting period be imposed by the new insurer for the same pre-existing condition.

39. Paul W. Newachek, "Improving Access to Health Care for Children, Youth, and Pregnant Women," *Pediatrics* 86 (1990): 626.

40. David S. Hilzenrath, "HMO's Save Money by Shifting Costs; Other Private Insurers Bear Burden of Inflated Hospital Rates," *Washington Post*, 6 June 1994, p. 1A.

41. H.R. 3600, 103d Cong., 1st sess. §1101 (1933).

42. 42 U.S. Code § 1395y(9)(7) (1990).

43. 42 U.S. Code § 1395y (1996) (regarding cosmetic services and routine dental and eye care). *Harris v. McRae*, 448 U.S. 297 (1980) (upholding the Hyde Amendment's prohibition on the use of federal funds for abortion, except in cases of rape, incest, or where the mother's life was in danger).

44. See Jessica Dunsay Silver, *From Baby Doe to Grandpa Doe: The Impact of the Federal Age Discrimination Act on the "Hidden" Rationing of Medical Care*, 37 Catholic U. L. Rev. 993 (1988).

45. 29 U.S. Code § 1185 (1996) (concerning group health plans organized under the Employee Retirement Income Security Act); 42 U.S. Code § 300gg-4 (1996) (concerning group health plans organized under the Public Health Service and within the reach of the federal government). For examples of state legislation, see Ill. Rev. Stat. ch. 5, para. 356r (1996); Ind. Code Ann. § 27-8-24-4 (Burns 1996); Md. Code Ann., Health-Gen. § 19-1305.4 (1995); Mo. Rev. Stat. § 376.1210 (1996); N.Y. Pub. Health Law § 2803-n (Consolidated 1996); Okla. Stat. tit. 35, § 6060.1 (1995).

46. 29 U.S. Code § 1001–1461 (1994).

47. See *Carparts Distrib. Ctr. v. Automotive Wholesaler's Assn. of New Eng.*, 37 F.3d 12 (1st Cir. 1994).

48. General Accounting Office, *Health Insurance: Coverage of Autologous Bone Marrow Transplantation for Breast Cancer*, GAO/HEHS-96-83 (Washington, D.C., 1996); William P. Peters and Mark C. Rogers, "Variation in Approval by Insurance Companies of Coverage for Autologous Bone Marrow Transplantation for Breast Cancer," *New England Journal of Medicine* 330 (1994): 473–477.

49. *Enrollees in Expanded Ore. Plan Exceed Projections by 27%*, 2 Health Care Policyy Rep. (BNA) d22 (8 Aug. 1994).

50. Eli Cailouto, et al., "How Restrictive Are Medicaid's Categorical Eligibility Requirements? A Look at Nine Southern States," *Inquiry* 29 (1992): 451.

51. S.B. 27, Or. Rev. Stat. §§ 414.025-414.750 (1989).

52. United States Congress, Office of Technology Assessment, *Evaluation of the Oregon Medicaid Proposal* (Washington, D.C.: U.S. Government Printing Office, 1992), 34.

53. John Kitzhaber, *The Oregon Basic Health Services Act* 5 (1989) (unpublished report).

54. OTA, "Oregon Medicaid Proposal," 39, n.2.

55. Ibid.

56. Golenski and Blum, *The Oregon Medicaid Priority-Setting Project* 12–16 (March 30, 1989).

57. OTA, "Oregon Medicaid Proposal," 64–78.

58. Ibid., 51.

59. Alexander M. Capron, "Oregon's Disability: Principles or Politics?" *Hastings Center Report* (November 1992): 18–20.

60. Letter from Louis Sullivan, Secretary of Health and Human Services, to Barbara Roberts, Governor of Oregon, 3 August 1992, reprinted in 9 Issues in Law & Med. 409 (1994).

61. Oregon Health Services Commission, *Prioritization of Health Services: A Report to the Governor and Legislator* (1993), 16–18.

62. Ibid.

63. "Oregon Carrying Out New State Health Care Plan: Questions Abound as First Phase Begins in Program to Cover All Residents," *Dallas Morning News*, 19 February 1994, p. 13A.

64. Robert H. Miller and Harold S. Luft, "Managed Care Plan Performance Since 1980: A Literature Analysis," *Journal of the American Medical Association* 271 (1994): 1512, 1516.

65. Ibid.

66. Section 1905(a)(4)(B) of the Social Security Act.

67. Omnibus Budget Reconciliation Act of 1989, Pulic Law 101-239 (1989).

68. See generally, Health Care Financing Administration, Department of Health and Human Services, "Medicaid Program: EPDST Services Defined, Proposed Rule," 58 *Federal Register* 51288 (1993).

69. Section 1903(i) of the Social Security Act.

70. See, for example, *Pereira v. Kozlowski*, 805 F. Supp. 361 (E.D. Va. 1992).

71. *Salgado v. Kirschner*, 878 P.2d 659 (1994).

72. Personal conversation with Ms. Sheri Dieterich, HIV Testing Coordinator of the Free Clinic, Cleveland, Ohio.

73. Duncan Neuhauser and Ann M. Lewicki, "What Do We Gain from the Sixth Stool Guaiac?" *New England Journal of Medicine* 5 (1975): 226–28.

74. Institute of Medicine, Committee on Assessing Genetic Risks, *Assessing Genetic Risks: Implications for Health and Social Policy* (Washington, D.C.: National Academy Press, 1994).

75. Gina Kolata, "Ticking Bomb: The Presence of a Breast Cancer Gene Creates Other Problems for Some Women," *Chicago Tribune*, 2 March 1997, p. 2; Peter J. Howe, "Evidence Cited on Cancer Gene: Researchers Say Defect Stops It from Warding Off Disease," *Boston Globe*, 24 January 1997, p. 26A.

76. H.R. 3600, §1114(a).

77. Institute of Medicine, *Assessing Genetic Risks*.

78. Ibid.

79. Henry J. Aaron and William Schwartz, *Rationing: The Painful Prescription* (Washington, D.C.: Brookings Institution, 1984).

80. See Or. Rev. Stat. § 414.725(7).

81. John K. Iglehart, "The American Health Care System: The End Stage Renal Disease Program, Health Policy Report," *New England Journal of Medicine* 328 (1993): 366–371.

82. Richard A. Rettig and Norman G. Levinsky, *Kidney Failure and the Federal Government* (Washington, D.C.: National Academy Press, 1991).

83. John K. Iglehart, "Transplantation: The Problem of Limited Resources," *New England Journal of Medicine* 309 (1983): 123.

84. Barri Bronston, "Working to Prevent Another Parent's Pain," *The Louisiana Times-Picayune,* 10 November 1996, p. 1E.

85. *Viveros v. State of Idaho Dep't. of Health and Welfare,* 889 P.2d 1104 (1995).

86. Telephone interview with Ira Greifer, M.D., Medical Director, National Kidney Foundation (28 June 1983).

87. Mark F. Anderson, *The Future of Organ Transplantation: From Where Will New Donors Come, to Whom Will the Organs Go?* 5 Health Matrix 249, 256 n.27 (1995); Arthur L. Caplan, "Obtaining and Allocating Organs for Transplantation," in *Human Organ Transplantation,* eds. Dale H. Cowan, et al. (Ann Arbor, Michigan: Health Administration Press, 1987), 55; American Medical Association, Council on Ethical and Judicial Affairs, "Ethical Considerations in the Allocation of Organs and Other Scarce Medical Resources Among Patients," *Arch. Intern. Med.* 155 (1995): 29, 31.

88. David C. Flower, *State Discretion in Funding Organ Transplants Under the Medicaid Program: Interpretive Guidelines in Determining the Scope of Mandated Coverage,* 79 Minn. L. Rev. 1233 (1995); Robin E. Margolis, *Medicaid Coverage of Organ Transplants,* 10 HealthSpan 27 (1993); David L. Weigert, *Tragic Choices: State Discretion Over Organ Transplantation Funding for Medicaid Recipients,* 89 Nw. U. L. Rev. 268 (1994).

CHAPTER 6: GENETIC TECHNOLOGIES AND THE CHALLENGE TO EQUALITY

1. See Richard A. Epstein, *Why Is Health Care Special?*, 40 U. Kan. L. Rev. 307 (1993).

2. Sally Lehrman, "Doctors Challenge Child Growth Hormone Tests," *The Phoenix Gazette,* 31 May 1993, p. 10B; Shari Roan, "Growth Drug Debate: What Is Too Short?" *Los Angeles Times,* 2 February 1993, p. 1E.

3. William Saffire, "Of I.Q. and Genes," *New York Times,* 20 October 1994, p. 27A; Ralph Heimer, "Smart Genes?" *New York Times,* 25 December 1994, p. 8; Myron A. Hofer, "Behind the Curve," *New York Times,* 26 December 1994, p. 39.

4. Norman Daniels, *Just Health Care* (Cambridge: Cambridge University Press, 1985).

5. Allen Buchanan, "The Right to a Decent Minimum of Health Care," in *Securing Access to Health Care,* ed. President's Commission for the Study of

Ethical Problems in Medicine and Biomedical and Behavioral Research (Washington, D.C.: U.S. Government Printing Office, 1983), 207–238.

6. Ibid., 216.

7. See, for example, Gerald R. Winslow, *Triage and Justice* (Berkeley: University of California Press, 1982), 63–86.

8. John Rawls, *A Theory of Justice* (Cambridge: Harvard University Press, 1971).

9. Ibid., 60.

10. See, for example, Thomas C. Shevory, *Applying Rawls to Medical Cases: An Investigation into the Usages of Analytical Philosophy,* 11 J. Health Pol., Pol'y & L. 749 (1986).

11. Larry R. Churchill, *Rationing Health Care in America: Perceptions and Principles of Justice* (Notre Dame, Indiana: University of Notre Dame Press, 1987), 49.

12. Robert Nozick, *Anarchy, State, and Utopia* (New York: Basic Books, 1974).

13. Paul T. Menzel, *Strong Medicine: The Ethical Rationing of Health Care* (New York: Oxford University Press, 1990).

14. Charles J. Dougherty, *American Health Care: Realities, Rights, & Reforms* (New York: Oxford University Press, 1988), 86.

15. Daniel Callahan, *What Kind of Life: The Limits of Medical Progress* (New York: Simon and Schuster, 1990); Daniel Callahan, *Settings Limits: Medical Goals in an Aging Society* (New York: Simon and Schuster, 1987).

16. Callahan, *What Kind of Life,* 255.

17. Ibid., 143–49.

18. Callahan, *Setting Limits.*

19. Robert H. Binstock and Stephen G. Post, *Too Old For Health Care? Controversies in Need, Law, Economics & Ethics* (Baltimore, Maryland: Johns Hopkins University Press, 1991); William J. Bricknell, "Set No Limits: A Rebuttal to Daniel Callahan's Proposal to Limit Health Care for the Elderly," *Journal of the American Medical Association* 269 (1993): 2148.

20. Maxwell J. Mehlman, *Age-Based Rationing and Technological Development,* 33 St. Louis U. L. Rev. 671 (1989).

21. Callahan, *What Kind of Life.*

22. Jane E. Brody, "Making It Work When Opposites Attract," *New York Times,* 23 November 1994, p. 7C.

23. Frank Parkin, *Class Inequality and Political Order: Social Stratification in Capitalist and Communist Societies* (New York: Praeger, 1971), 48.

24. David B. Grusky and Azumi Ann Takata, "Social Stratification," in *Encyclopedia of Sociology,* ed. Edgar F. Borgatta and Marie L. Borgatta (New York: Macmillan, 1992).

25. John H. Schaar, *Legitimacy in the Modern State* (New Brunswick, New Jersey: Transaction Books, 1981), 195 (emphasis added).

26. Plato, *The Republic,* trans. and ed. H.D.P. Lee (Harmondsworth, England: Penguin, 1955).

27. Ibid., 285.

28. Ibid., 280.

29. Pitirim A. Sorokin, *Social Mobility* (New York: Harper, 1927), 533.

30. Parkin, *Class Inequality,* 50.

31. Ibid., 57.

32. Joseph A. Califano, Jr., *Radical Surgery: What's Next for America's Health Care* (New York: Times Books, 1995); Willard Gaylin, "Faulty Diagnosis," *New York Times,* 12 June 1994,p. 4A.

33. Willard G. Manning, et al., "The Taxes of Sin: Do Smokers and Drinkers Pay Their Way?" *Journal of the American Medical Association* 261 (1989): 1604.

34. Howard Leichter, "Public Policy and the British Experience," *Hastings Center Report,* (October 1982): 32.

35. Frederick A. Connell and Jane Huntington, "For Every Dollar Spent—The Cost-Savings Argument for Prenatal Care," *New England Journal of Medicine* 331 (1994): 1303–1307.

36. See Chapter Five.

37. See Callahan, *Setting Limits.*

38. See discussion of coverage of genetic services under national health reform in Chapter Five.

39. Maxwell J. Mehlman, *The Patient-Physician Relationship in an Era of Scarce Resources: Is There a Duty to Treat?* 25 Conn. L. Rev. 349 (1993).

40. For a description of the interplay of these factors, see Richard Rettig, *The Policy Debate on Patient Care Financing for Victims of End-Stage Renal Disease,* 40 Law & Contemp. Probs. 201, 202 (1976).

CHAPTER 7: RESPONDING TO THE CHALLENGE

1. See "Rationing Called Dilemma for the '80s," *American Medical News,* 12 November 1982, 2.

2. Lester M. Salamon, *America's Nonprofit Sector: A Primer* (New York: Foundation Center, 1992); Bradford H. Gray, *Why Nonprofits? Hospitals and the Future of American Health Care,* 8 Exempt Org. Tax Rev. 729 (1992).

3. American Association of Fund-Raising Counsel, *Giving USA* (New York: American Association of Fund-Raising Counsel, 1994), 10.

4. See, for example, Karen Hsu, "Breda: Small Iowa Town That's a Different Breed," *Des Moines Register,* 24 July 1994, p. 1R.

5. Edmond N. Cahn, *The Moral Decision* (Bloomington: Indiana University Press, 1955), 71.

6. Amy Gutmann, "For and Against Equal Access to Health Care," in *Securing Access to Health Care,* ed. President's Commission for the Study of Ethical Problems in Medicine and Biomedical and Behavioral Research (Washington, D.C.: U.S. Government Printing Office, 1983), 53.

7. Charles J. Dougherty, *American Health Care: Realities, Rights & Reforms* (New York: Oxford University Press, 1988), 54.

8. Gutmann, "For and Aganist Equal Access," 51–66.

9. Ibid., 52.

10. Ibid., 55.

11. Dougherty, *op. cit.,* at 55.

12. See Note, *Scarce Medical Resources,* 69 Colum. L. Rev. 620, 653 (1969).

13. "Withdrawal of Interim NIH Guidelines for the Support and Conduct of Therapeutic Human Fetal Tissue Transplantation Research," 58 *Federal Register* 45, 495 (1993).

14. National Institutes of Health, Human Embryo Research Panel, *The National Institutes of Health Report of the Human Embryo Research Panel* (Bethesda: National Institutes of Health, 1994).

15. Paul Berg, et al., "Asilomar Conference on Recombinant DNA Molecules," *Science* 188 (1975): 991.

16. Donald S. Fredrickson, "Asilomar and Recombinant DNA: The End of the Beginning," in *Biomedical Politics,* ed. Kathi E. Hanna (Washington, D.C.: National Academy Press, 1991), 258.

17. 45 C.F.R. § 46.101, et sec.

18. 21 C.F.R. § 50.23.

19. American Medical Association, Council on Ethical and Judicial Affairs, "Ethical Issues Related to Prenatal Genetic Testing," *Archives of Family Medicine* 3 (1994): 633, 640–41.

20. See Daniel Callahan, *Setting Limits: Medical Goals in an Aging Society* (New York: Simon and Schuster, 1987).

21. See, for example, Jon S. Batterman, *Brother Can You Spare a Drug?* 19 Hofstra L. Rev. 191 (1990).

22. Zad Leavy and Jerome Kummer, *Criminal Abortion: Human Hardship and Unyielding Laws,* 35 S. Cal. L. Rev. 123 (1962).

23. 21 U.S.C. § 841.

24. See Doug Bandow, *War on Drugs or War on America?* 1991 Stan. L. & Pol'y Rev. 242.

25. See *Zemel v. Rusk,* 381 U.S. 1 (1965) (upholding restrictions on travel to Cuba).

26. See *Haig v. Agee,* 453 U.S. 280 (1981) (upholding revocation of passport).

27. See Immigration and Nationality Act of 1952, 8 U.S. Code § 1185 (1976) (amended 1978).

28. 42 C.F.R. § 34.2(b) (1997).

29. 19 U.S. Code § 482 (1996).

30. See, for example, *U.S. v. Himmelwright,* 551 F.2d 991 (5th Cir. 1977).

31. International Emergency Economic Powers Act, 50 U.S. Code §§ 1701-06 (1996).

32. Currency and Foreign Transactions Reporting Act, 31 U.S. Code § 5316 (1996).

33. 31 U.S. Code § 5317 (1996).

34. See, for example, *California Bakers Ass'n v. Shultz,* 416 U.S. 21 (1974).

35. Gutmann, "For and Against Equal Access," 53.

36. Personal communication with Dr. Peter Whitehouse.

37. See, for example, Ronald P. Keeven, "Pros and Cons of Gambling Amendment: Money Used for Legal Betting Drains Resources of the Poor," *St. Louis Post-Dispatch* 27 March 1994, p. 3(B).

38. Ibid.

39. University of Chicago Press, 1992.

40. *Holmes v. New York City Hous. Auth.*, 398 F.2d 262 (2d Cir. 1968).

41. *Hornsby v. Allen*, 330 F.2d 55 (5th Cir. 1964).

42. Deborah S. Pinkney, "Firm Faces Legal Flak Over Drug Monitoring Rules," *American Medical News*, 2 November 1990, p. 33.

43. Susan K. Miller, "MS Drug Shortage Prompts Patient Lottery," *New Scientist* 139 (1980): 8.

44. Ibid.

45. *U.S. v. Holmes*, 26 F. Cas. 360 (E.D. Pa. 1842) (No. 15,383).

46. Ibid., 367.

47. H.R. 3600, 103d Cong., 1st sess. § 7201(a) (1993).

Index

abortion, 28, 33, 50, 66, 84, 118
  as alternative to gene therapy, 84
access to genetic technologies. *See also*
    equality; handicapping,
    genetic; lotteries, genetic
  ability to pay, role of in, 85–87, 105
  black markets, 118
  genetic aristocracy, and creation
    of, 99, 102
  government payment for, 107–108.
    See also access to genetic technol-
    ogies, role of insurance in,
    Medicare, Medicaid
  information, role of in, 81–82
  insurance, role of in, 61–87
    Blue Cross/Blue Shield, 65
    capitation, 82
    copayments and deductibles,
      69
    cost shifting, 62
    DRG's, 65
    ERISA, 66, 112
    experimental treatments, insur-
      ance coverage of, 67
    gag clauses, 81
    gene therapy, insurance cover-
      age of, 82–84
    genetic counseling, insurance
      coverage of, 80–81
    genetic enhancement, insur-
      ance coverage of, 84–85
    genetic testing, insurance
      coverage of, 77–82
    HMO's. *See* access to genetic
      technologies, role of insur-
      ance in, managed care plans
    managed care plans, 63, 73–76
    Medicaid, 62, 66, 68, 69, 76, 86,
      112, 116. *See also* access to
      genetic technologies, role of
      insurance in, Oregon Medic-
      aid program
    medical savings accounts, 88
    Medicare, 62, 65, 112, 116
    Oregon Medicaid program, 69–
      73, 86–87, 110
    preexisting conditions, 62, 68
    private insurance, 66
    state mandates, 66, 112
  market, role of in, 88–89
  national health reform, effect on,
    64–65, 87, 128
  philanthropy, role in, 110–112
  prohibiting, 112–121. *See also*
    genetic research, prohibiting
    financial restrictions, 120–121
    travel restrictions, 119–120
  social mobility, role in promoting,
    61, 101–102
  willingness to pay, role of in, 89–
    90
ADA deficiency, 7
adenine. *See* nucleotides
adoption, 28
adrenal hyperplasia, congenital,
    27
adult onsent polycystic kidney
    disease, 22, 24, 38
affirmative action programs. *See*
    handicapping, genetic
aging. *See* genetic engineering,
    physical traits, longevity